2007 年 3 月在被授予博士学位仪式上

北海道的冬天

博士研究生毕业典礼

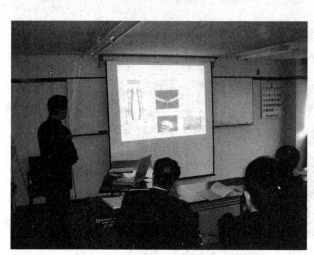

2007 年 1 月在博士论文答辩会上

2007 年与妻子在日本北海道

留学生交流会

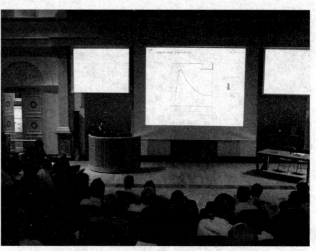

2010 年参加在比利时举行的国际学术会议

2008 年参加在英国举行的国际学术会议

为小学生做体验实验

在日本大学教学中

耐磨合金开发及磨损性能评价

新巴雅尔　著

内蒙古科学技术出版社

图书在版编目（CIP）数据

耐磨合金开发及磨损性能评价 / 新巴雅尔著. — 赤峰：内蒙古科学技术出版社，2017.10（2022.1重印）

ISBN 978-7-5380-2860-7

Ⅰ. ①耐…　Ⅱ. ①新…　Ⅲ. ①耐磨合金—研究　Ⅳ. ①TG135

中国版本图书馆CIP数据核字（2017）第254209号

耐磨合金开发及磨损性能评价

作　　者：新巴雅尔
责任编辑：那　明
封面设计：永　胜
出版发行：内蒙古科学技术出版社
地　　址：赤峰市红山区哈达街南一段4号
网　　址：www.nm-kj.cn
邮购电话：0476-5888903
排版印刷：三河市华东印刷有限公司
字　　数：350千
开　　本：889mm×1194mm　1/16
印　　张：12
版　　次：2017年10月第1版
印　　次：2022年1月第3次印刷
书　　号：ISBN 978-7-5380-2860-7
定　　价：58.00元

序

在以航空发动机等为首的重型机械行业中，多数设备是在磨损工作环境下运行的，其零部件耐磨性能的好坏左右着整个产品的性能和可信赖度。尤其是零部件的耐磨性及其磨损寿命直接影响其使用成本，故从经济角度分析，磨损现象成为一项急需解决的技术难题。目前在国内外，从优化机器的结构特性、改善使用条件及提高使用材料的耐磨性等方面进行过许多研究，并取得了一定的成效。但是，磨损仍然是"搞不懂"的东西。也就是说，实际发生的磨损现象往往较复杂，定量的理解和分析较困难，实验室里的评价与实际现象的不符等一直以来是磨损不可解的重要原因，虽然新设备、新材料的应用能在一定程度上减少耐磨材料的消耗量。我国每年在煤炭、冶金、水泥、电力、建筑材料、农机等部门消耗金属耐磨材料及能源所造成的经济损失是惊人的。据不完全统计，我国每年煤炭系统因磨损失效而需要更换的机械设备价值高达10亿~20亿元，煤炭开采企业为此付出了沉重的代价。同时，因为机器设备更换停机而造成的间接经济损失更是无法计算。还有，近10年的统计表明，我国仅在水泥企业，各类金属耐磨耐热材料消耗总量就达40万吨以上。因此，在研制开发具有优异耐磨性的新型耐磨材料的同时，一定要研究其严酷环境下的磨损机理和预测耐磨材料寿命，这不仅是提高经济效益和社会效益的一项重大课题，而且对地区经济发展具有迫切性和现实意义。

本书简要介绍了著者十几年来从事耐磨材料的开发，材料与结构的磨损行为与磨损检测、评估，材料的使用寿命预测等方面研究所取得的成果。

本书由8章组成。

第1章，阐述了目前关于冲蚀磨损问题，指出了冲蚀磨损在工业现场中的重要性，同时也指出了冲蚀磨损特性评价的重要研究意义，最后指出本研究的目的。

第2章，研究了工业现场的重要结构材料——铸铁冲蚀磨损，采用不规则形状的硅砂作为冲蚀颗粒，靶材诸如弯管、阀门中广泛使用的球墨铸铁和灰口铸铁。本章中，证实了基本的冲蚀磨损试验方法和磨损特性评价方法。另外，还探讨了该方法的有效性及实验结果整理方法。

第3章，作为冲蚀磨损实验中使用的靶材，利用含有球状碳化物的三种不同基体的铸铁（Spheroidal Carbides cast Iron: SCI）和对比材料的高铬铸铁（WCI），在第2章中确立并验证过的试验方法来进行冲蚀磨损实验。结果表明，球状碳化物铸铁表现出非常优异的耐冲蚀磨损性能，比常规耐磨材料WCI高1/4~1/3，更重要的是几乎没有显示冲蚀角度依赖性。

第4章，为澄清球状碳化物铸铁的冲蚀磨损机理，利用金相显微镜和扫描电镜观察磨损表面附近的连续观察图和表面SEM形貌图。此外，利用双试样连续截面观察方法，在金相显微镜下每磨损一定时间后观察磨损形貌，理解冲蚀磨损详细过程。此外，通过磨损过程的模型化试图阐明铸铁球状碳化物的冲蚀磨损机理。

第5章，以球状碳化物铸铁为研究材料，研究分析其磨粒磨损性能。

第6章，利用铸渗方法实现在球状碳化物的表面层上进行表面改性，开发出球状碳化钨（WC）增强的复合材料,试图增强球状碳化物铸铁抗冲蚀磨损性能。

　　第7章,通过结合消失模铸造与镶铸法开发出廉价的钢铁基复合材料,研究其磨粒磨损性能。

　　第8章,总结本研究中取得的成果。

目　录

2

第1章　冲蚀磨损

1.1　序言

　　磨损是指摩擦副对偶面在相对滑动（接触）过程中，表面材料逐渐损耗的一种现象。相互作用的表面间一定会发生磨损。摩擦副材料表面在受损的同时，在外力的作用下继续恶化。除自适应磨损以外，大多数磨损是有害的，它会增加表面损耗，引起不必要的位移、表面精度的下降，导致大的振动或负载，进而促使疲劳磨损。在大型复杂装置中，残存的微小的损伤将可能引起毁灭性故障。[1]

　　磨损习惯上分为以下几个类型：

　　1. 磨粒磨损（磨料磨损）　由于硬质物料或突出物与表面相互摩擦使材料发生损耗的现象。

　　2. 腐蚀磨损　环境介质与材料表面发生的化学或电化学反应，伴随机械作用使材料损失的现象。

　　3. 黏着磨损　在黏着力的作用下使材料在两个表面发生迁移的现象。

　　4. 疲劳磨损　由于交变应力的不断作用使材料疲劳脱落的现象。

　　按照不同的分类方式，磨粒磨损又可以分成不同的磨损形式，其中按磨料对材料的力学作用特点可分为以下3个类型：

　　（1）凿削磨粒磨损　在较严重冲击载荷下的磨粒磨损，例如颚式破碎机颚板上发生的磨损等。

　　（2）研磨磨粒磨损　在高应力下的磨粒磨损，例如球磨机磨球和衬板的磨损等。

　　（3）刮伤磨粒磨损　在低应力下的磨粒磨损，也称作冲蚀磨损或冲刷磨损，例如渣浆泵叶轮和护套的磨损等。

　　在实际生产中，磨损往往并不是以单一类型存在的，而是几种磨损形式同时存在并相互影响，但其中总有一种占主导地位。所以工程人员分析材料磨损状况时，一般采取下面的步骤：首先要弄清主导磨损类型和磨损机理，再结合材料的使用环境及条件，得出材料磨损的原因，并提出具体的防范措施及改进方案。

　　本书中，重点讨论冲蚀磨损现象。液体或固体以松散的小颗粒在流体或气体载体的作用下，按一定的速度或角度对材料表面进行冲击所造成的一种材料损耗现象或过程，称之为冲蚀磨损（erosion, erosive wear）或干砂冲刷（sand erosion）。冲蚀磨损现象广泛存在于输送装置中的配管弯头、阀门、涡轮叶片、直升机叶片及风扇等部件中，已成为造成重大事故和经济损失的重要原因之一。[2-5]

　　下面是关于冲蚀磨损的一个实例。在炼钢厂二次精炼及溶液还原等设备中，微小的粉煤、矿石粉等的颗粒从通风口（羽口）吹入时，管道内壁的弯曲部位常会发生冲蚀磨损引起的损伤破坏。Fig.1-1-1中表示炼钢高炉中的喷碳装置（PCI: pulverized coal injection system）中管道内壁及其他输送物料管道的气动运输装置中发生的冲

1

蚀磨损现象。这种管道内壁一旦磨损出现孔洞，在气体和颗粒的喷刷作用下将可能引起严重的事故。

如此冲蚀磨损是发生在管道内壁，而从外部无法判断其磨损程度，故现阶段只能依靠定期维修时，采取更换管道或在磨损部位进行堆焊增厚等临时措施。所以，开发耐磨材料、推测材料余寿命，从而把事故降至最低是亟待解决的重大课题。

一提耐磨材料，尤其是合金耐磨材料，就会想到以高铬铸铁为中心的多种钢铁合金材料、复合材料等，因为这些材料一度称霸于耐磨材料领域，直到现在也继续发挥其应有的作用。然而，随着新型材料的开发和不断更新，加上工作环境的苛刻度增加，也给耐磨材料提出更高的要求，这意味着更适合现状的新型耐磨材料的开发是必然的趋势。比如，高铬铸铁具有优良的耐磨性和耐腐蚀性，因而在耐磨材料中得到广泛应用。高铬铸铁等白口铸铁系列材料应用在水力发电机的主要铸造零件上，如内衬板、导向叶片等，用来抵抗冲蚀磨损。还有，在许多干砂冲刷磨损情况的地方，也用高铬铸铁作为抗冲蚀磨损材料。但是，该类合金的最大缺点是，其组织中大量存在树枝状或板条状M_7C_3或$M_{23}C_6$碳化物，在其受到冲击载荷时产生严重的应力集中现象，使其冲击值急剧下降，最终以脆性断裂告终。因此，本书中主要讨论作为耐磨材料，如何改善其内部微观组织结构，开发出抗冲击能力好、耐磨性强的铸造合金。

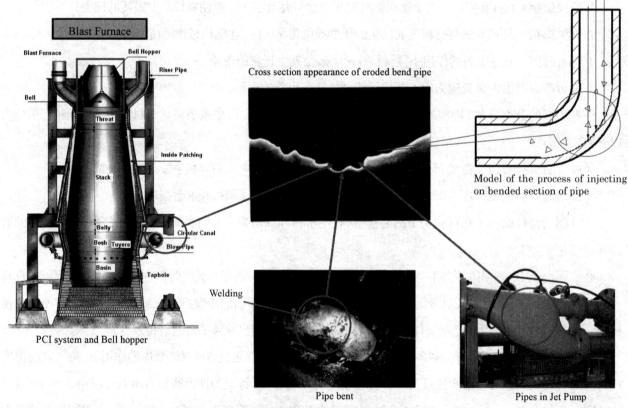

Fig.1-1-1　Erosion phenomena in some industrial sites

1.2　关于冲蚀磨损的研究现状

冲蚀磨损的研究工作，始于20世纪中叶。随着大型工业的迅速发展，得到众多工程人员和研究人员的重视，从20世纪70年代开始迅速发展起来。从此，关于冲蚀磨损方面的研究论文、防范措施、解决方案已经逐步浮出水面。下面概括这些研究成果，主要体现在以下两个方面：

（1）从目前的研究成果看，研究者进行过许多磨损试验，来解释靶材的冲蚀磨损机理。Fig.1-2-1表示为典型的磨损量和冲蚀角度的关系图。图中表示，如铝合金、铜及低碳钢等韧性材料在20°~30°低冲蚀角度时磨损量最大；与其相反，陶瓷、玻璃等脆性材料在80°~90°高冲蚀角度时达到最大的磨损量。很明显，如上面所述，材料在冲蚀磨损过程中存在冲蚀角度的依赖性。

作为基础研究，1960年美国加州大学的Finnie等人[6]，对于韧性材料受到不定型颗粒的冲击时冲蚀磨损进行了详细描述。他利用二元切削理论，建立了基于切削模型的预测公式，大致解释了冲蚀颗粒的冲蚀角度和冲蚀速度对其冲蚀磨损的影响规律。但是，该公式只能推测出与磨损量达到最大值的20°~30°附近的实测值相近的值，而达到40°以上的高角度时，与实测值的相差变得很大。针对这种情况，Finnie的解释是，在高冲蚀角度的情况下，随着磨损的进行，由于材料表面变粗糙并形成凹凸不平的表面，实际的冲蚀角度将会变得与起初的冲蚀角度不同。但是，Finnie的公式中，冲蚀角度为90°时，磨损量变为零，这种现象目前得不到解释。此外，也可以看出，冲蚀角度为90°时的磨损量不能和冲蚀速度的二次方成正比例关系。

Bitter[7, 8]针对磨损量预测进行了综合的、系统的研究，发展了Finnie的理论。Bitter认为，超过某一界限磨损条件发生变化，如下解释了冲蚀磨损中冲蚀角度依赖性问题：低的冲蚀角度磨损时，发生切削型磨损；当变为高角度时，磨损形式变为主要以变形磨损为主。

接着，Neilson and Gilchrist等人[9]对Bitter的结论进行了简化。

还有，Tilly[10-13]引入了初始磨损对冲蚀角度的影响，对韧性材料的冲蚀磨损提出了二阶段磨损机理。Hutchings等人[14, 15]针对大倾斜角和球形颗粒情况的冲蚀磨损机理，提出了耕犁（ploughing）机理。其定性说明如下：①颗粒在材料表面的一次冲击，只能使压痕尖端的突起部分变形，但不能从表面被除去。②接下来的颗粒冲击使突起状的变形部位优先被（切削）脱落。

综上所述，以上基础性研究奠定了目前关于冲蚀磨损机理方面的理论依据。

（2）从上述可以看出，到目前为止，关于冲蚀磨损方面的众多基础研究，都不能够充分说明无论是韧性材料还是脆性材料的磨损过程、磨损机理。更值得注意的是，直到20世纪80年代几乎没有关于实际工况中应用最广泛的材质由韧性到脆性大范围变动的钢铁材料的冲蚀磨损研究。基于此观点，仅仅Finnie, Bitter等人的研究成果不能完全对应各种材质的磨损情况。这说明，对实际应用广泛的钢铁材料的冲蚀磨损基本特性的把握是必要的。

关于实际应用广泛的钢铁材料的冲蚀磨损特性，日本清水一道等人[2-5, 16, 17]的研究最为突出。清水等人在冲蚀磨损特性评价过程中，使用吸引式喷砂型磨损机，利用球形及不定型的钢球作为冲蚀颗粒进行了冲蚀磨损实验。实验中所涉及的材质为软钢/碳钢SS400, S25C, S45C（日本国内的钢号，相当于国标中Q235, 25#, 45#），铁素体基和珠光体基球墨铸铁，不锈钢及高强韧性的奥贝铸铁（ADI）等是实际应用最为广泛的钢铁材料。

Fig.1-2-2, Fig.1-2-3, Fig.1-2-4中分别列出碳钢、铸铁、ADI的冲蚀角度与损伤速度（erosion rate）之间的关系。

软钢/碳钢在冲蚀角度20°~30°的低角度附近达到最大磨损量，在60°附近一度减少，然后在90°附近取第二次峰值。然而，铸铁的冲蚀角度依赖性与钢大不相同，不同种类的铸铁的最大损伤速度相差很大，而且在20°～30°的低角度时损伤速度变小。随着冲蚀角的增加损伤速度也会增加，损伤速度在冲蚀角度大约为60°时变为最大。此后，随着冲蚀角度的增大，损伤速度降低。在ADI中，冲蚀角度在40°～60°附近达到最大值，而两端显示较低的值。特别应该注意的结果是，在ADI的损伤速度与软钢相比非常低，表现出优异的抗冲蚀磨损性能。理由是，在其基体中含有的残留奥氏体，在颗粒冲击作用下由应变诱导相变转变成马氏体，从而提高了硬度。从钢铁的研究中也可以看出，对于抗冲蚀磨损材料来说，对于不同的颗粒，均存在相应的冲蚀角度依赖性。

清水一道确立了磨损过程连续观察法，阐明了球铁组织中的球状石墨很容易成为磨损的起点，在磨损过程中也会促使材料的进一步磨损。也进一步提倡预测冲蚀磨损寿命时，应充分考虑在当前公开发表的磨损推导公式的基础上，还要考虑对冲蚀磨损作用产生巨大影响的因子中加工硬化后的硬度变化等，以此来推导磨损关系更为合适。

此外，20世纪90年代以后，关于冲蚀磨损的研究，从基础研究转入以上述基础理论为基础的耐冲蚀磨损材料开发研究为主。比如说，许多研究人员[18-22]把研究重点放在复合材料、高温合金及表面涂层等的抗冲蚀磨损特性方面，也取得不少可喜的成果。但是，还没有得到一种能够直接应用到现场的具有优异的抗冲蚀磨损性能的材料。

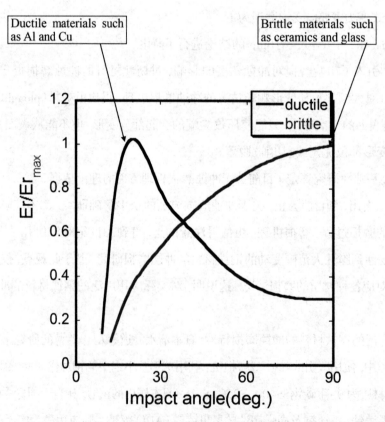

Fig. 1-2-1　Erosion rate vs. Impact angle for ductile and brittle materials by solid particles

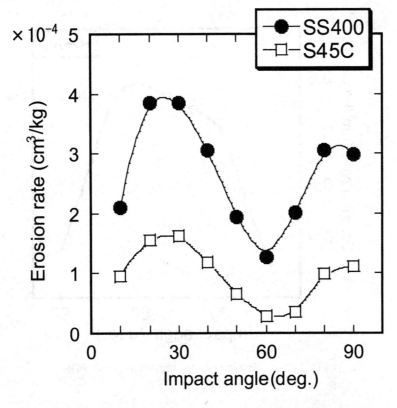

Fig. 1-2-2　Erosion rate vs. Impact angle for mild steel

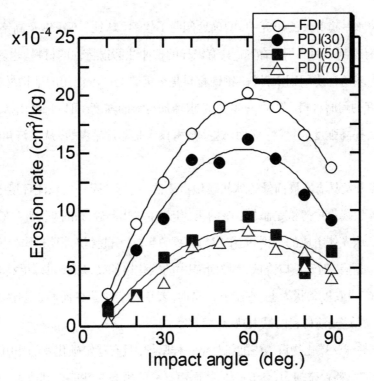

Fig. 1-2-3　Erosion rate vs. Impact angle for cast irons

5

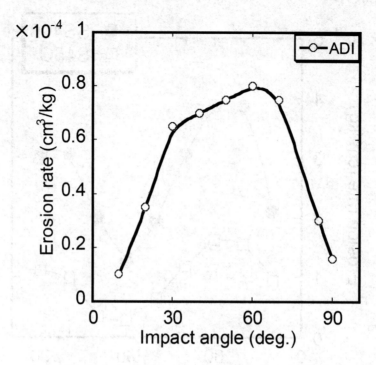

Fig. 1-2-4   Erosion rate vs. Impact angle for ADI

1.3   本书的编写目的

6

目前，在日本的清水一道等人的研究中，明确地阐明了软钢、球铁、不锈钢及高铬铸铁等的冲蚀磨损特性[2-5, 16, 17]，但是还没有得到一种使用寿命长，具有优异的抗冲蚀磨损特性的材料。也就是说，到目前为止，作为冲蚀磨损的靶材，许多场合使用钢铁材料。钢铁材料是今天最为广泛使用的结构材料。靶材等的选择是希望找到比目前材料更好的耐磨材料。然而，一味地追求增加材料的强度，可以选择由钛、钨等构成的表面增强材料及进行HIP处理后提高强度的材料。然而，这些技术投入是非常昂贵的，从经济角度考虑，它不可能作为实际管道材料来使用。

所以，本书中重点开发具有发展前景的球状碳化物铸铁，进行了固—气双流环境下的砂砾冲蚀引起的磨损特性评价。开发的球状碳化物铸铁是添加10%的钒，组织中析出高硬度的碳化钒（VC）的一种铸铁。灰口铸铁因为是由片状石墨和珠光体组成的，片状石墨成为磨损的起点，引起材料损伤，作为抗冲蚀磨损材料使用时不能得到良好的效果。还有，对于白口铸铁，因为组织中的碳化铬的硬度高，可以提高其耐磨耗性，但是由于这些碳化物的形状以带状或板条状为主，在受到冲击时，易引起应力集中而冲击韧性下降，从而变得容易断裂而缩短部件使用寿命，这是该材料的最大缺点。

另一方面，添加钒的球状碳化物铸铁中的钒与铬一样坚硬，且有稳定碳化物的作用。提高材料力学性能，通过析出球形VC以分散由颗粒碰撞引起的应力，从而使材料的耐磨损性能得到改善。在本研究中，主要目的是开发具有不同基体结构的球状碳化物铸铁，进行冲蚀磨损实验研究。

在以前的磨损实验中，使用过钢砂进行过冲蚀磨损实验。实验中利用的钢砂一般市场都能买得到，其平均粒径为660 μm，维氏硬度为420 HV。靶材有碳钢和ADI。碳钢的硬度约为600 HV，ADI的初始硬度为350

HV, 但磨损表面检测到大约硬度提高到700 HV, 那是由于固体颗粒的碰撞引起加工硬化的结果。

从Fig.1-2-3, Fig.1-2-4中可以总结出, 靶材的硬度比冲蚀颗粒的硬度越高, 其耐磨性就越好。所以, 从磨损量与固体颗粒的硬度方面考虑, 如冲蚀颗粒的硬度高于靶材的硬度时, 可以推断磨损量必然会大幅度增加。

在本书中, 通过冲蚀磨损实验, 使用比钢砂硬的硅砂作为固体颗粒。硅砂的平均粒径为410 μm, 维氏硬度为1030 HV。利用硅砂的理由如上文所述, 硅砂的硬度比高铬铸铁（硬度为600~730 HV）、球状碳化物铸铁（硬度为450~730 HV）和软钢（硬度约为150 HV）要高。另外, 也能够对应工业现场的情况, 如高炉中的喷碳装置中被喷入的不定型的颗粒, 对这种耐冲蚀磨损特性的评价是可以胜任的。

基于上述, 在本书中, 通过了解各种球状碳化物铸铁的冲蚀磨损性能, 试图解释其冲蚀磨损机理。根据以往的研究, 一般使用的靶材都存在冲蚀角度依赖性, 即材料的冲蚀磨损特性取决于冲蚀角度的变动。然而, 特别是在管道输送装置中, 当粉体被输送时, 在所有的角度上都有可能发生碰撞, 所以耐磨材料最好是无冲蚀角度依赖性。鉴于这一事实, 掌握开发的球状碳化物铸铁的耐冲蚀磨损性能及其磨损机理是非常有必要的。

参考文献

[1] HUTCHINGS I M. Tribology, Friction and Wear of Engineering Materials [M]. Oxford UK: Butterworth-Heinemann, 1992: 1-2.

[2] K.SHIMIZU T N, S.DOI. Basic Study on the Erosive Wear of Austempered Ductile Iron [J]. Transactions of AFS, 101 (93-78), 1993: 225-9.

[3] KAZUMICHI SHIMIZU T N E. fundamental study on erosive wear of austempered ductile iron [J]. proceeding of the third east Asian international foundry symposium, 1992:

[4] SHIMIZU K, NOGUCHI T. Erosion characteristics of ductile iron with various matrix structures [J]. Wear, 176 (2), 1994: 255-60.

[5] SHIMIZU K, NOGUCHI T. Erosion Characteristics of Ductile Iron with Various Matrix Structures [J]. THE JOURNAL OF THE JAPAN FOUNDRYMEN'S SOCIETY, 66 (7), 1994: 489-94.

[6] FINNIE I. Erosion of surfaces by solid particles [J]. Wear, 3 (2), 1960: 87-103.

[7] BITTER J G A. A study of erosion phenomena part I [J]. Wear, 6 (1), 1963: 5-21.

[8] BITTER J G A. A study of erosion phenomena [J]. Wear, 6 (3), 1963: 169-90.

[9] NEILSON J H, GILCHRIST A. Erosion by a stream of solid particles [J]. Wear, 11 (2), 1968: 111-22.

[10] TILLY G P. Erosion caused by airborne particles [J]. Wear, 14 (1), 1969: 63-79.

[11] TILLY G P. Sand erosion of metals and plastics: A brief review [J]. Wear, 14 (4), 1969: 241-8.

[12] TILLY G P, SAGE W. The interaction of particle and material behaviour in erosion processes [J]. Wear, 16 (6), 1970: 447-65.

[13] TILLY G P. A two stage mechanism of ductile erosion [J]. Wear, 23 (1), 1973: 87–96.

[14] HUTCHINGS I M, WINTER R E. Particle erosion of ductile metals: A mechanism of material removal [J]. Wear, 27 (1), 1974: 121–8.

[15] HUTCHINGS I M. Prediction of the resistance of metals to erosion by solid particles [J]. Wear, 35 (2), 1975: 371–4.

[16] KAZUMICHI SHIMIZU T N A S A. Erosion Characteristics of Ductile Iron with Various Matrix Structures [J]. Trans of the Japan foundrymen's society, 13 (1994)

[17] SHIMIZU K, NOGUCHI T, KAMADA T, et al. Progress of erosive wear in spheroidal graphite cast iron [J]. Wear, 198 (1), 1996: 150–5.

[18] SHIMIZU K, NOGUCHI T, KAMADA T, et al. Basic Study on Erosion of Ductile Iron [J]. Advanced Materials Research, 594 (4), 1997.

[19] DIVAKAR M, AGARWAL V K, SINGH S N. Effect of the material surface hardness on the erosion of AISI316 [J]. Wear, 259 (1–6), 2005: 110–7.

[20] HUSSAINOVA I. Microstructure and erosive wear in ceramic–based composites [J]. Wear, 258 (1–4), 2005: 357–65.

[21] WHEELER D W, WOOD R J K. Erosion of hard surface coatings for use in offshore gate valves [J]. Wear, 258 (1–4), 2005: 526–36.

[22] MISHRA S B, PRAKASH S, CHANDRA K. Studies on erosion behaviour of plasma sprayed coatings on a Ni–based superalloy [J]. Wear, 260 (4–5), 2006: 422–32.

8

第2章　一般铸铁的冲蚀磨损特性

2.1　引言

铸铁是工业中重要的结构材料,其广泛应用于作为受冲蚀磨损的部件。铸铁的强度、硬度、延展性和韧性随着其碳含量、热处理工艺和微观组织的变化而发生变化。清水一道等人[1]针对以SS400为代表的具有广泛应用的常规管材和球墨铸铁等,利用球形钢砂做了许多冲蚀实验。本章中,采用无定型石英砂对广泛应用于弯管和阀门的球墨铸铁和灰铸铁进行冲蚀磨损实验,确定了基本冲蚀磨损实验的重复性,为磨损特性评价奠定了基础,并讨论其有效性,提供了整理实验结果的方法。

2.2　基本实验方法

2.2.1　冲蚀磨损试验机及试验方法

冲蚀磨损试验采用了商用的吸引式喷砂机。其示意图和实物照片分别如Fig.2-2-1和Fig.2-2-2所示。本设备利用高压气流将固体颗粒高速喷射到样品表面,进行冲蚀磨损试验。利用商用的喷砂机有如下优点:

（1）应用广,不受地域和环境限制,容易入手。

（2）降低不同研究者之间的试验误差。

由于冲蚀磨损性能受到很多未知因素的影响而发生变化,所以统一试验设备规格对耐磨试验的研究极为重要。本试验采用的喷砂机参数如下:

空气压力: 5 kg/cm²;

流量: 0.31 m³/min;

风机功率: 2.2 kW;

空气流速的测试和调整:利用压力计进行;

喷砂喷嘴直径: 6 mm。

粒子随着气流从喷嘴喷出,但是粒子速度和空气流速是不同的。清水等人通过采用高速相机测试球形钢砂的冲蚀速度时发现,作为标准的直径为0.66 mm的（质量约1 mg）球形钢砂的速度为空气流速的15%,约为21 m/s,同样,不同粒径钢砂的冲蚀速度也可以控制。因此,在本研究中,即使利用不定型石英砂,在不测试粒子的冲蚀速度的情况下,也可以控制冲蚀速度。此外,虽然粒子粒径随着冲蚀时间变化而变化,但是在一定范围内可以控制冲击速度。

    试样为50 mm×50 mm×10 mm的平板。冲蚀磨损时粒子冲击试样的角度很重要,因此,需要定制冲蚀角度可以在0~90°范围内随意变化的样品台,并把样品固定在样品台上。喷嘴和试样中心的距离为50 mm,并且此距离不随冲蚀角度的变化而变化。如Fig.2-2-3所示,从喷嘴喷出的粒子到试样的张开角约为6°。

    冲蚀磨损试验均是在室温下进行的,其空气流速(V)约为100 m/s。冲蚀粒子喷射量,使用不定型石英砂时设定为4~5 g/s,钢砂为37.0 g/s,冲蚀角度为30°,60°,90°。

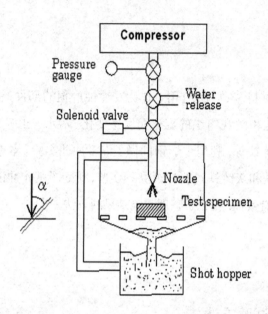

Fig. 2-2-1   Schematic diagram of the blast machine

Fig. 2-2-2　Blast machine and its testing stage

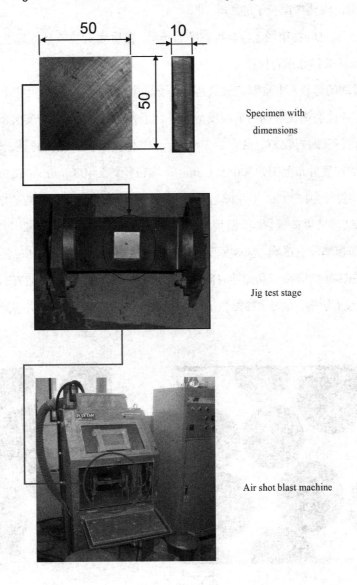

Specimen with dimensions

Jig test stage

Air shot blast machine

Fig. 2-2-3　Procedure before doing erosion test

2.2.2　固体颗粒

固体粒子，即冲蚀颗粒使用了平均粒径为660 μm，硬度为440 HV的钢砂或平均粒径为408 μm，硬度为1030 HV的不定型石英砂。钢砂和石英砂的宏观照片如Fig.2-2-4所示。其中钢砂作为对比物质，而使用不定型石英砂的理由为：

（1）经常发生冲蚀磨损的高炉的喷煤装置所吸入的粒子是粒径在100~500 μm范围内的不定型粉尘和固体。

（2）前人的研究结果表明[2, 3]，冲蚀颗粒的晶粒大于100 μm时，粒子的影响变得次要；另外，冲蚀颗粒的尺寸超过一定值时还需要考虑粒子的影响。[1]

（3）由于使用常见的在市场上容易得到的石英砂，在后续实验或其他地方做实验时容易买到同种规格的石英砂，从而保证实验结果的可比性。

随着试验时间的延长，冲蚀颗粒被破碎，粒子尺寸变小，当粒子的尺寸小于临界尺寸时，会影响冲蚀磨损。因此，我们对冲蚀磨损试验前后的粒度分布进行了表征。Fig.2-2-5为实验前的新石英砂的粒度分布图，实验3 600sec.后的石英砂粒度分布如Fig.2-2-6所示。靶材为球状碳化物铸铁，测定颗粒样品数量为130个。实验前，新石英砂的粒度分布范围为300~500 μm，平均粒径约为408 μm；而冲蚀3 600sec.后的粒度分布范围为100~180 μm，平均粒径为122 μm，即冲蚀3 600sec.后平均粒径减少到新砂的约1/4。因此，长时间使用石英砂可能会对冲蚀磨损试验结果造成影响。但是，根据Tilly[3]的试验结果（Fig.2-2-7），冲蚀粒子的粒径超过100 μm时，对靶材的损伤速率没有影响，与本实验的冲蚀颗粒的变化范围一致。Fig.2-2-8论述了本研究中使用的实验材料之一——SCI-W的磨损量和冲蚀时间的关系。从图中可知，磨损量随着时间的延长直线增加，表明粒径的变化对磨损量没有影响，即表示冲蚀粒子的粒径在允许的变化范围内。本实验每进行3 600 sec.试验后换新的石英砂。

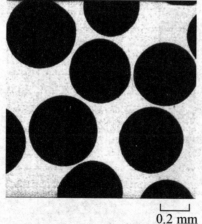

0.2 mm

Spherical steel grits with average diameter of 660μm and with hardness of 420HV

0.2 mm

Irregularly shaped silica sand with average diameter of 408μm and hardness of 1030HV

Fig.2-2-4　Macrostructure of impact particles

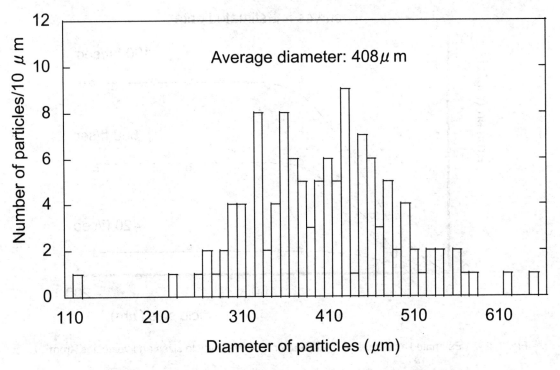

Fig.2-2-5　Particles size distribution of silica sand before erosion test

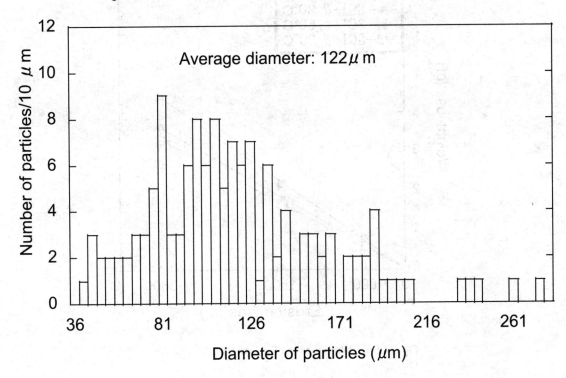

Fig. 2-2-6　Particles size distribution of silica sand after erosion test

13

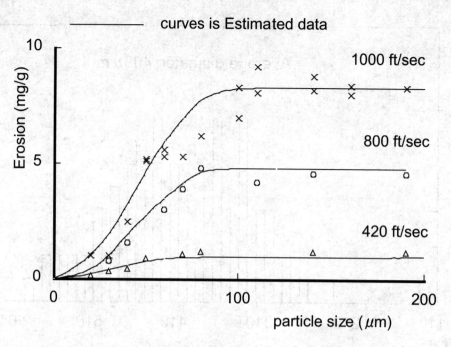

Fig. 2-2-7　Estimated and experimental data for different particle sizes and velocities (from [4])

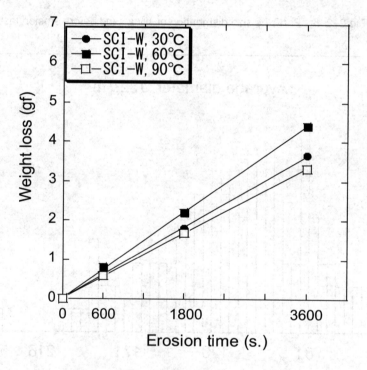

Fig. 2-2-8　Wear loss vs. Erosion time for SCI-W

2.2.3　实验试样

实验材料采用了常见的灰铸铁（FC200）和球墨铸铁（FCD400）。具体的化学成分和金相组织分别如Table2-2-1, Table2-2-2和Fig.2-2-9所示。磨损试样的外观和尺寸如Fig.2-2-10所示, 试样为50 mm× 50 mm×10 mm的平板。为了统一磨损初期状态, 所有试样进行统一研磨。以下均对研磨后的试样进行了实验。

Table2-2-1　Chemical compositions of FC200 and FCD400

Specimens	C	Si	Mn	Ni	Cr	V
FC200	3.24	1.79	0.26	0.06	—	—
FCD400	3.74	2.16	0.30	0.02	—	—

Table 2-2-2　Vickers hardness of specimens

	FC200	FCD400
Vickers hardness	256 HV	221 HV

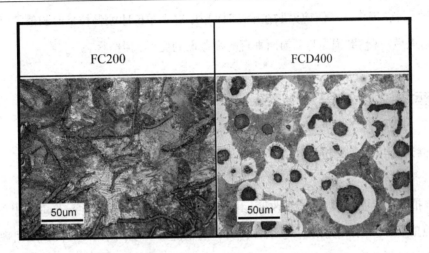

Fig.2-2-9　Microstructure of FC200 and FCD400

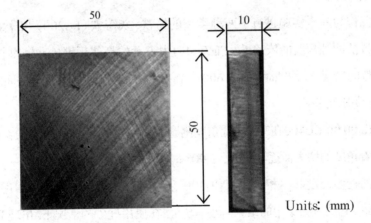

Fig. 2-2-10　Dimensions of the test piece

2.3　实验结果整理方法

首先采用精度为0.01 mg的天平对磨损质量进行称量, 然后通过除以试样密度计算出磨损体积。在对比具有不同密度样品的冲蚀磨损性能时, 对比体积减小量更为合适(而不是重量减小量)。因此, 采用靶材的平均密度对损伤速率(Erosion rate)进行如下定义:

$$\text{单位时间内的体积损伤量 }(cm^3/s) = \frac{\text{单位时间内的质量减少量 }(g/s)}{\text{材料的平均密度 }(g/cm^3)}$$

$$\text{损伤速率 }(cm^3/kg) = \frac{\text{单位时间内的体积损伤量 }(cm^3/s)}{\text{单位时间内粉体的喷射量 }(kg/s)}$$

Finnie, Erosion of Surfaces by Solid Particles, Wear, 3, p87-103, 1960

其中, 用单位时间内的冲蚀颗粒量除体积磨损量的理由如下:

认为一定时间内的磨损量应该取决于该时间内冲蚀颗粒的总量。如果冲蚀颗粒的量小, 试样表面的损伤也小, 反之亦然。因此, 损伤速率值应该取在单位时间内, 这样单位冲蚀颗粒量时的值才更合理。

磨损试验结束后, 观察磨损面与磨损面垂直的横截面的变形、损伤状态。

2.4 冲蚀磨损试验结果及考察

2.4.1 铸铁材料的冲蚀磨损特性

以前的实验报道中表明, 材料的类型改变, 其冲蚀角度依赖性也随之改变。同时, 冲蚀磨损过程中每个材料都有弱的冲蚀角度和强的冲蚀角度。总结各种材料的情况时发现, 低角度和高角度出现峰值的情况, 或者中间角度出现的特征角度依赖性的情况最多, 所以只要在30°, 60°, 90° 三个代表性的角度上执行实验就能够掌握材料冲蚀角度的依赖性。

在本研究中, 首先, 作为基本实验进行了两种类型的铸铁、灰铸铁 (FC200) 和球墨铸铁 (FCD400) 的冲蚀磨损实验来证实其特征冲蚀角度的依赖性。Fig.2-4-1中表示的是使用钢砂冲蚀颗粒, Fig.2-4-2中表示的是使用石英砂冲蚀颗粒的情况下, 冲蚀时间为3 600 s后时的冲蚀速度和损伤速率之间的关系。图中横坐标表示冲蚀角度, 纵坐标表示损伤速率。

对钢砂的情况, FC200和FCD400在30° 的低角度侧磨损量小。随着角度增大, 损伤速率也增加, 到60° 时达到最大值。然后, 随着角度的增大而逐渐变小, 在90° 时损伤速率最小。

FC200和FCD400的初始硬度基本相当, 但FCD400显示出较低的磨损值。其原因可认为FCD400组织中的球状石墨会分散由碰撞引起的应力集中的结果。另一方面, 在使用硅砂的情况下, 出现30° 的低角度取最大损伤速率, 在60° 时取最小值, 最后随冲蚀角度的增加, 损伤速率逐渐增加到到90° 时出现另一个峰值的特性曲线, 并且磨损量急剧增大, 其原因是石英砂的硬度远远超出靶材的硬度, 导致了剧烈磨损。

2.4.2 对冲蚀磨损面的观察

为了把上述结果与磨损表面变化状态相结合进行分析, 也为了更进一步阐明各试样的磨损机理, 冲蚀磨损实验后对其冲蚀磨损表面采取宏观观察。Fig.2-4-3中是表示使用钢砂冲蚀颗粒, Fig.2-4-4中是表示使用石英砂冲蚀颗粒情况下的冲蚀磨损表面宏观照片。清水一道等人研究结果表明, 在S45C试样中, 使用钢砂作为冲蚀颗粒时，出现与冲蚀方向垂直的鲜明条纹。然而，在本研究中，虽然在试样FC200和FCD400中，观察到了像在S45C中那样清楚的条纹，但在冲蚀颗粒为石英砂的任何冲蚀角度的情况下，冲蚀磨损表面就看不到像在S45C使用钢砂冲蚀粒子的情况下出现的条纹。

另外，对于这两种冲蚀颗粒，分别对比冲蚀角度为30°，60°，90°的表面情况时发现，任何材料在低角度时受损区域宽，并朝向冲蚀方向扩张；高角度时变得接近于圆形，并且损伤面积较小。下一节中，通过详细观察磨损表面附近的横截面及磨损表面SEM，分析出现这种现象的原因。

2.4.3　对磨损表面附近的横截面组织的观察

为了进一步分析磨耗表面的变化和由于冲蚀角度的不同产生损伤速率的差异，在3 600 s试验完成后，把试样沿着正中间切开，然后研磨、抛光该切断面，观察了磨损表面附近的横截面微观结构。Fig.2-4-5分别表示试样FCD400在使用钢砂粒和干砂时，冲蚀角度分别为30°和60°时截面结构状态。我们已经得知损伤量的值表现为特征冲蚀角度依赖性的事实，与此对应的磨损表面形貌也出现了明显的差异。在角度为30°时，通过颗粒的冲击，磨损表面附近发生塑性变形，其组织流向冲蚀方向的下游侧，使表面在下游形成伸出的突起物，并在颗粒的持续冲击下继续延伸。通过随后的粒子碰撞，使伸出部位破裂、分离和脱落，如此重复进行磨损。另一方面，在冲蚀角度为60°时，虽然同30°类似地形成向下游流动的突起，但是其大小与30°相比要小。可以看出，如果使用钢砂，冲击形成的突起部的伸长非常大，肉眼也能观察到形成的条纹。对于石英砂，由冲击造成的突起，也通过沙砾的切割作用被切削掉后，其伸长变短，所以肉眼看不出条纹。但是，两者的磨损形貌可以理解成是一样的。

2.4.4　对磨损面的微观观察

冲蚀磨损实验完成后，为了进一步详细观察磨损表面的形态，利用扫描电子显微镜（Scanning Electron Microscope），观察了磨损表面形貌。Fig.2-4-6中表示了FC200和FCD400的冲蚀磨损表面的SEM照片。图中表示的是使用石英砂（为充实颗粒），冲蚀时间3 600 s后的表面形貌。从SEM照片可以看出，在30°角度的冲蚀磨损中，被观察到由于颗粒的冲击塑性变形的金属表面被切割的痕迹。在90°冲蚀磨损时，能够确认，通过颗粒的冲击后，表面组织发生塑性变形，形成大的突起和压痕。在60°的情况下，同时可以观察到，塑性流动的金属表面被切割掉的现象和表面组织产生塑性变形而形成小的突起和压痕的现象。上述现象与磨损横截面观察得到的结果一致。

因此，通过磨损表面的宏观观察、磨损表面附近的横截面微观观察、磨损表面的SEM观察，可以理解，冲蚀颗粒不论是在钢砂还是在石英砂的情况下，磨损表面的磨损形态（类型）是一样的。另外，在工业现场中，高炉的粉煤喷射系统（喷煤系统）内部被吹入的颗粒物或输送系统中被输送的物体都是以不规则形状的固体颗粒为主，所以为了揭示这种环境下发生的冲蚀磨损特性，在模拟实验（实验室的小型实验）中不是利用球形颗粒，而是希望利用不规则形状的颗粒。基于上述，本研究执行了干砂冲蚀磨损试验来评估磨损环境。

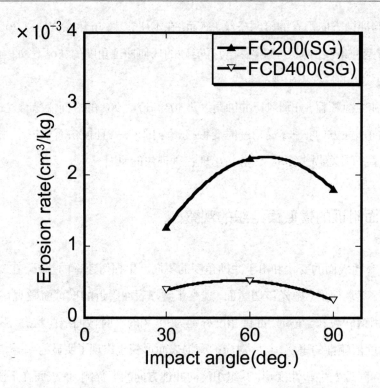

Fig. 2-4-1　Erosion rate vs. Impact angle for FC200 and FCD400 by steel grits

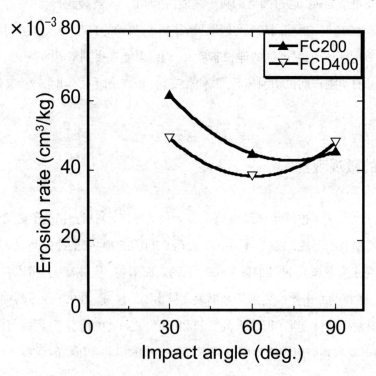

Fig. 2-4-2　Erosion rate vs. Impact angle for FC200 and FCD400 by sand particles

18

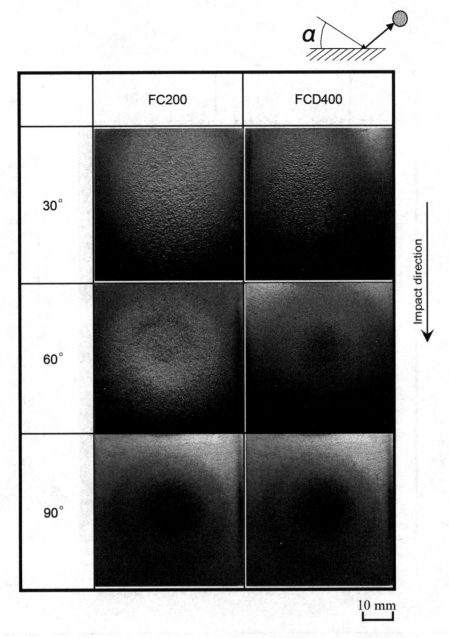

Fig. 2-4-3 Eroded surface in FC200 and FCD400 by steel grits

19

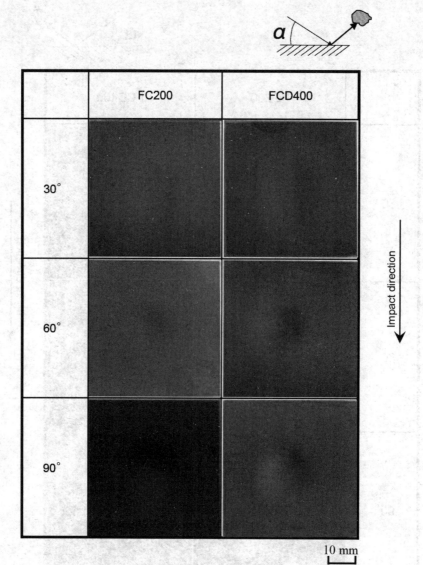

Fig. 2-4-4   Eroded surface of specimens by sand particles

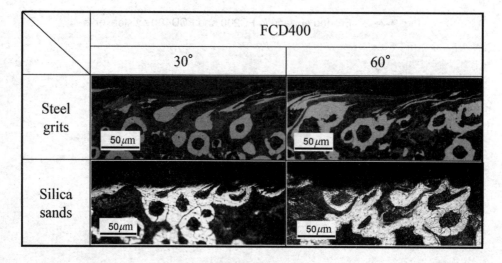

Fig. 2-4-5   Vertical section observation of FCD400 in the condition of
steel grits and sand particles at 30° and 60°

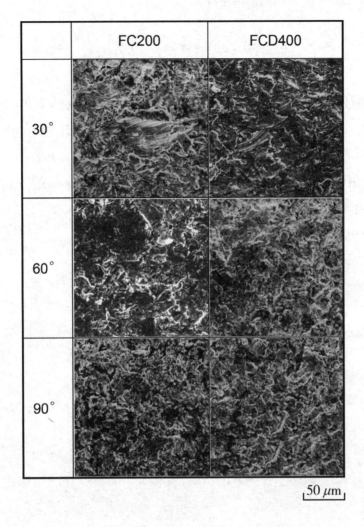

Fig. 2-4-6　SEM observation of FC200 and FCD400

2.5　小结

本章首先确认冲蚀磨损实验方法（1）的再现性，通过其冲蚀磨损试验的基本方法和磨损评价法，利用吸引式喷砂机来进行冲蚀磨损实验。以FC200，FCD400为靶材，使用钢砂粒和石英砂为冲蚀颗粒，进行预备实验。通过此种方法，第一，确认了冲蚀磨损实验方法的再现性；第二，确定了利用石英砂的颗粒交换时间。其次，使用日本清水一道的研究中所提倡的损伤速率（Erosion rate）来总结冲蚀磨损的结果。[1]最后，通过进行表面宏观观察、磨损表面的SEM观察，以及磨损表面附近的横截面微观观察等，分析靶材的冲蚀角度依赖性。其结果如下：

（1）冲蚀颗粒不论是钢砂还是石英砂，同样发生切割磨损和变形磨损。

（2）干砂冲蚀磨损试验的目的，是为了获得与高炉的粉煤喷射系统（喷煤系统）或运输系统的实际情况最接近的实验数据。

参考文献

[1] 清水一道. 鉄鋼材料のエロージョン摩耗特性 [D]. Hokkaido University, 2001.

[2] MISRA A, FINNIE I. On the size effect in abrasive and erosive wear [J]. Wear, 65（3）, 1981: 359–73.

[3] TILLY G P. A two stage mechanism of ductile erosion [J]. Wear, 23（1）, 1973: 87–96.

[4] WINTER R E, HUTCHINGS I M. Solid particle erosion studies using single angular particles [J]. Wear, 29（2）, 1974: 181–94.

22

第3章　各种球状碳化物铸铁的冲蚀磨损特性评价

3.1　序言

在第2章中，针对广泛应用于弯管、阀门等的普通铸铁，如灰铸铁（FC200）和球墨铸铁（FCD400），用球形钢砂和不定型石英砂进行了冲蚀磨损实验，提出了基本的实验方法和实验结果整理方法，从而确认了清水一道等人[1]的实验方法和实验结果整理法的可重复性。截至目前的研究表明，材料有各自的冲蚀角度依赖性，从延展性材料如铝到脆性材料如玻璃，都具有不同的冲蚀角度依赖性。另一方面，在输送系统中，管道的弯曲部位发生的冲蚀磨损，因为冲蚀颗粒从所有的角度接触到管道内壁，所以有必要改善这种配管用材料的冲蚀角度依赖性。

到目前，清水一道等人[2-6]的研究虽然已经揭示了软钢、球铁、不锈钢和高铬铸铁等的冲蚀磨损性能，但是还没有得到一种其冲蚀角度依赖性得到改善的以及具有长寿命的优良的耐冲蚀、耐磨损材料。

因此，在本章中，针对所开发的几种球状碳化物铸铁，进行了固—气环境下的干砂冲蚀磨损特性评价，对特有的冲蚀磨损特性进行了详细的研究。

球状碳化物铸铁是通过基体中添加钒，约10mass%而析出高硬度的球状碳化钒（VC）的铸铁。认为和铬一样，钒具有稳定碳化物的效能，并通过析出球状的VC，将颗粒碰撞产生的应力分散，能够改进材料的冲蚀耐磨性能。相比较，灰口铸铁因为微观组织是片状石墨和珠光体，其石墨部分成为磨损的起点，很容易扩展至损伤，所以不能成为良好的抗冲蚀磨损材料。而高铬铸铁的情况是，硬度虽然高，但在其基体组织中包含的板条状$M_{23}C_6$等碳化物容易导致应力集中，表现出低的断裂韧性等特点。

3.2　球状碳化物铸铁的冲蚀磨损特性评价

3.2.1　试样材料及试验方法

冲蚀磨损实验中使用的试样为基体组织中含有球状碳化物的三种球状碳化物铸铁（Spheroidal Carbides cast Iron：以下简称SCI）。它们分别是，具有不同基体组织的不锈钢系球状碳化物铸铁（以下称之为SCI-VCrNi），高Mn系球状碳化物铸铁（以下称之为SCI-Mn）和白口系球状碳化物铸铁（以下称之为SCI-W）。SCI-VCrNi和SCI-Mn分别是不锈钢系列和高锰系列的铸铁，共同点是基体都是奥氏体组织，VC为球状。SCI-W是白口铸铁系列的铸铁，基体组织为硬度较高的贝氏体组织。对比材料选用为，高铬铸铁（含Cr量为29mass%，以下简称29Cr),它具有耐磨性好,在严酷的磨损环境下优于钢铁系列材料,广泛用在水力发电站等的重要部件。其主要化学成分为Table3-2-1中,初始硬度为Table3-2-2中,以及试样的金相组织为Fig.3-2-1中分别所示。

Table3-2-1   Chemical composition (mass %) of SCI series and 29Cr

	C	Si	Mn	Cr	V	Ni	others
SCI-VCrNi	2.84	0.95	0.61	17.3	**9.34**	9.2	—
SCI-VMn	2.92	0.57	12.9	—	**11.9**	—	—
SCI-W	2.79	0.96	0.54	—	**12.7**	3.06	Mo
29Cr	2.79	0.89	0.91	**28.69**		0.025	P,S

Table3-2-2   Initial Vickers hardness of SCI series and 29Cr

SCI-VCrNi	SCI-VMn	SCI- W	29Cr
399HV	504 HV	418 HV	610 HV
Micro-Vickers hardness of vanadium is: ≈2300 HV			

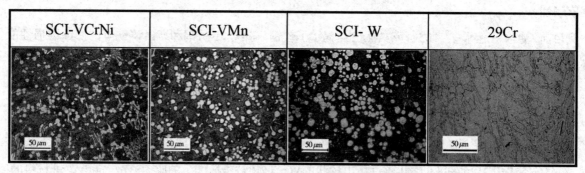

Fig. 3-2-1   Microstructure of SCIs and 29Cr

下面详细介绍球状碳化物铸铁。本研究中所涉及的实验材料,即球状碳化物铸铁(其英文定义为Spheroidal Carbide Cast Iron; 以下简称SCI)是一种组织中加入10mass%钒后析出高硬度的球状碳化钒(VC; 2 300 HV)的铸铁。碳化钒的硬度非常高,维氏硬度可达到2 300 HV左右。熔炼球状碳化物铸铁的原材料被加热到1 750 ℃时,已析出的VC将球化;但在1 600 ℃时,球化不均匀或出现不同形状的VC,所以要想得到完全均匀的球化VC,需要较高的工艺技术和温度管理方法。

通过碳化物的球化能够使耐磨性得到提高,其强韧性与球墨铸铁、耐腐蚀性与不锈钢相媲美,也被认为在腐蚀环境下表现出优异的耐磨损性。通过析出球状的碳化钒(VC),应力被分散,以改善材料的耐磨性。钒也和铬一样,具有稳定碳化物的效能,所以也被认为能提高材料的抗冲蚀磨损性能。

关于SCI系列铸铁中析出球状VC碳化物的研究,有日本的山本悟、西内滋典等人[7、8]的研究。在他们的研究中提出了两种生成球状碳化物的说法,一是偏析球化机理,二是气泡球化机理。偏析球化机理可以解释为在高温下变成球形的现象,但无法解释碳化钒当温度降低时由球形变成不规则形貌的现象。而在气泡球化机理中,从量子力学的基本概念出发,从结构的角度更合理地说明了在高温下,将把V作为气泡源成分进行添加,熔融金属中气体(氢气)的气泡,使球形空间内部优先析出球状碳化钒,而在低温时,因不存在气泡而不能成为球状的现象。

球状碳化物铸铁是综合考虑了以往的铸铁的优点和缺点开发出的一种崭新的材质,所以该材料与其他的铸铁和复合材料相比具有非常优异的力学性能。

(1)是具有优异的耐磨性、耐腐蚀性的金属材料,超越现有的非金属陶瓷和异种复合材料所具有的功能。

(2)可以回收再利用,克服了常规的非金属陶瓷和异种复合材料等使用后的磨损剩余材料的再利用困难和不可能再利用的弱点。

(3)球状碳化物铸铁的最大特性是,将铸件组织结构中的碳化物按照球形的VC(碳化钒)形式均匀地分散到金属基体中。所以这种结构被认为与高铬铸铁所含的枝晶状碳化物组织的脆性相比,具有分散应力集中的特性,从而具有进一步提高耐冲击性的效果。

下面介绍冲蚀磨损试验方法。磨损试验机是商用的吸引式喷砂机。使用的试样尺寸为50 mm×50 mm×10 mm的平板试样,将试样固定在试验机中的试样台上,将冲蚀角度改变为30°,60°,90°,进行了磨损试验。固体颗粒硬度为1030 HV、平均粒径为408 μm的石英砂。每个实验中使用2 kg的砂子,冲蚀速度为100 m/s,单位冲蚀颗粒量为4~5 g/s,实验时间定为3 600 s,所有的实验是在常温下进行的。实验前后利用电子天平(灵敏度为0.01 mg)称量试样的重量,计算出重量差来评价磨损特性。

3.2.2　高 V–Cr–N 球状碳化物铸铁的冲蚀磨损特性

Fig. 3–2–2表示SCI–VCrNi的冲蚀时间为3 600 s时的冲蚀角度和损伤速率之间的关系。图中还一并表示出上一章中的FC200, FCD400结果和对比材料29Cr的结果。

SCI–VCrNi球状碳化物铸铁中,冲蚀角度为30° 时,其损伤速率达到最大值(1.259 5×10^{-2}cm³/kg);在30° ~60° 之间,一时减少;在60° 时取最小值(1.049 1×10^{-2}cm³/kg),然后,损伤速率随着角度的增大而增大;在90° 时再次出现峰值(1.406 4×10^{-2}cm³/kg)。而最小和最大之间的差只有3.373×10^{-3}cm³/kg。另一方面,29Cr的情况是,低角度侧30° 的损伤速率少(3.264×10^{-2}cm³/kg);随着角度增加,损伤速率继续增加,最大损伤速率(3.47×10^{-2}cm³/kg)出现在60° ;然后,随着角度的增大损伤速率减少,在90° 时取最小值(3.044×10^{-2}cm³/kg)。SCI–VCrNi的最大损伤速率(1.46×10^{-2}cm³/kg)与29Cr最大损伤速率(3.47×10^{-2}cm³/kg)进行比较,前者大约只有后者的1/3。而且其最大值与最小值之差也比起29Cr的情况小很多,可以看出大约只有1/4。

同时也可以确定,29Cr中出现的冲蚀角度依赖性较大,而SCI–VCrNi表现出显著抑制了冲蚀角度依赖性。FC200和FCD400的情况很显然,出现明显的冲蚀角度依赖性,低冲蚀角度30° 时其损伤速率达到最大,然后,损伤速率随着角度的增大而减少,在 60° 时取最小值,然后,损伤速率随着角度的增大而增大。

比较损伤速率得知,FC200的最高值(6.173×10^{-2}cm³/kg)约为SCI–VCrNi的最高值的6倍,FCD400的最高值(4.93×10^{-2}cm³/kg)约为SCI–VCrNi最高值的5倍。

这些展现完全不同的冲蚀角度依赖性的原因,应在于它们的基体组织中的第二相的性质(分布、形状、大小、硬度等)不同所致。分布在FCD400的基体重球状石墨在灰铸铁中变成片状时,虽然冲蚀角度依赖性不受影响,但是损伤速率却上升。还有,虽然为板条状形状,但比石墨更硬的碳化铬的析出明显降低了损伤速率。冲蚀角度依赖性表现为与含石墨的铸铁相反的特性。从板条状碳化物变为球状碳化物(VC)时,损伤速率减少了一个数量级,而且,冲蚀角度依赖性也消除了。

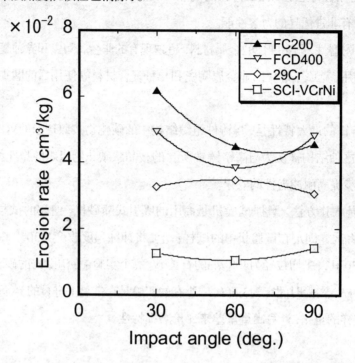

Fig. 3-2-2　Erosion rate vs. Impact angle in FC200, FCD400, 29Cr and SCI-VCrNi by sand particles

3.2.3　高 Mn 系球状碳化物铸铁的冲蚀磨损特性

冲蚀磨损试验中的试验材料为:灰铸铁(FC200),球墨铸铁(FCD400),高铬铸铁(29%的Cr)。基体中含有球状碳化物的高Mn球状碳化物铸铁(Spheroidal Carbides cast Iron: SCI-VMn)。

首先,冲蚀颗粒选为球形钢砂,试验材料分别选为FC200,FCD400和SCI-VMn进行了冲蚀磨损实验。选用钢砂作为冲蚀颗粒的理由是,形状简单统一,直径和密度一致,既可以对比之前的实验结果,也可以容易与随后的实验结果相比较。Fig.3-2-3中表示各试样在冲蚀时间为3 600 s时的冲蚀角度和损伤速率之间的关系。用钢砂做实验时,SCI-VMn的磨损很小,表现出良好的耐冲蚀磨损性能,FC200和FCD400出现较大的冲蚀磨损。这两种试样都在30°的低角度侧磨损量小,随着角度增加,损伤速度继续增加,到达60°时取最大值,然后,损伤速率随角度增加而减少,在90°变得最小。

使用比钢砂硬度高、不定型的石英硅砂,与前面相同的试样材料执行了冲蚀磨损试验的结果在Fig.3-2-4中表示。当使用高硬度、不定型的石英砂时,FC200,FCD400和29Cr中产生了严重磨损,甚至SCI-VMn受到磨损,但是没有出现清晰的峰。从这里可以判断出,当冲蚀粒子改变时,不仅磨损显著变化,而且材料的冲蚀角度依赖性也受到影响。这与Shimizu等人[1]和Finnie等人[9]的研究相一致。SCI-VMn的损伤速率是FC200的1/8,是FCD400的1/6,是常被广泛用作耐磨材料的29Cr的1/4,也看不出其冲蚀角度依赖性。

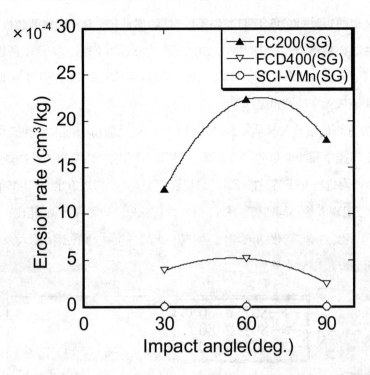

Fig. 3-2-3 Erosion rate vs. Impact angle in FC200, FCD400 and SCI-VMn by steel grits

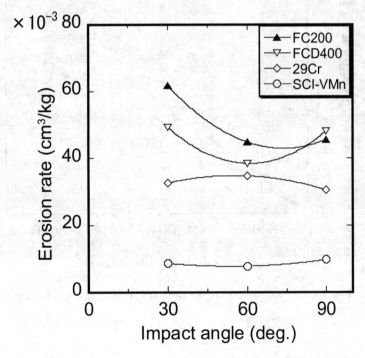

Fig. 3-2-4 Erosion rate vs. Impact angle in FC200, FCD400, 29Cr
and SCI-VMn by sand particles

3.2.4 白口铸铁系球状碳化物铸铁的冲蚀磨损特性

　　Fig.3-2-5中表示在以使用石英砂为冲蚀颗粒的情况下, 试样SCI-W中冲蚀角度分别为30°, 60°, 90° 时, 冲蚀磨损量和冲蚀磨损时间之间的关系。从图中可以看出, 对于3个冲蚀角度, 如果冲蚀磨损中的初期磨损

（磨损达到稳定状态之前的短时间的磨合期）忽略不计时，磨损量几乎随冲蚀时间线性增加。比较30°，60°，90° 时试验材料的冲蚀磨损量得知，在30° 和90° 的情况下，直线的斜率几乎一致，所以认为磨损量是大致相同的；而在60° 的情况下，直线的斜率稍微变得陡一些，磨损量有稍大的倾向，但其差别微小，可以认为SCI-W 的冲蚀磨损特征是几乎不依赖于冲蚀角度。

SCI-W使用石英砂进行的实验与其他材料的实验结果（按照损伤速率来整理的结果）表示在Fig.3-2-6 中。Fig.3-2-6中表示损伤速率和冲蚀角之间的关系。虽然高铬铸铁29Cr常用作抗冲蚀磨损材料，其损伤速率是SCI-W的2倍，并在60° 附近出现峰值，在两侧上的损伤速率均低。可以看出，SCI-W的损伤速率与冲蚀角没有多大关系，在所有角度出现大约相同的值。另外，与两种铸铁FC200和FCD400相比，大大降低了损伤速率，大约降低到原来的1/6。事实表明，碳化物的球化，使SCI-W的对抗冲蚀磨损性能大大增强，同时冲蚀角度依赖性也得到很好的改善。

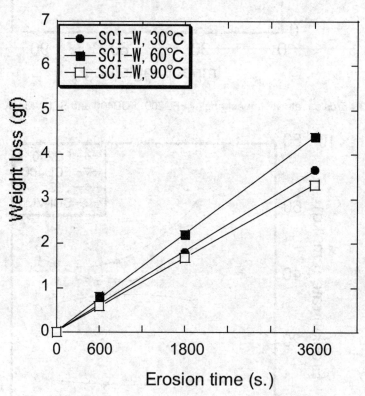

Fig. 3-2-5　Weight loss vs. Erosion time of SCI in 1h of blasting

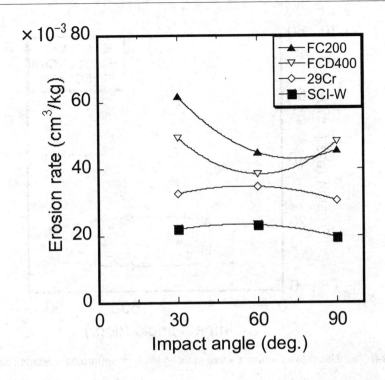

Fig. 3-2-6　Erosion rate vs. Impact angle in FC200，FCD400，29Crand SCI-W by sand particles

3.3　各种球状碳化物铸铁的冲蚀角度依赖性的比较研究

在前一节，进行了三种类型的球状碳化物铸铁的冲蚀磨损实验，并与其他结构材料的实验结果分别进行了对比。在本节中，总结该三种类型的试样冲蚀磨损实验结果。Fig.3-3-1中表示该材料的冲蚀角度依赖性的对比结果。Fig.3-3-2中表示所有试样的冲蚀磨损实验结果。

在Fig.3-3-1中，三种试样对比，SCI-VMn的值为最小，其次为SCI-VCrNi，而SCI-W表现出相对高的损伤速度。可见，三种类型的球状碳化物铸铁的损伤速度都很低，表现出优异的抗冲蚀磨损性能。

如Fig.3-3-2所示，先分析球状碳化物铸铁的情况，在所有的试验材料中，其损伤速度显著降低，显示出良好的耐磨损性，而且这些球状碳化物铸铁的损伤速度，几乎不随冲蚀角度的变化而变化，损伤速度在30°到90°范围内的变化只有 $(0.194 \sim 0.356\,4) \times 10^{-2}$ cm³/kg，表现出无冲蚀角度依赖性。在另一方面，到目前为止，其他所有试样中，都出现冲蚀角度依赖性，每个试样都表现出具有特征角度依赖性。而这些冲蚀角度依赖性在现实中已成为该材料的弱点，限制其使用。与此相反，球状碳化物铸铁表现出不依赖于固体颗粒冲蚀角度变化的特性。

综上所述，可以认为，通过在铸铁组织析出球状碳化物，可以分散由冲蚀颗粒冲击带来的应力集中，并吸收能量，这样由冲击形成的突起变小，脱落也延缓，结果抗冲蚀磨损性得到大幅度提高。但是，为了阐明其不依赖于冲蚀角度的特性，还需要进一步分析研究。

球状碳化物铸铁具有优异的耐冲蚀磨损性能的直接原因之一是试验后磨损表面硬度变化。

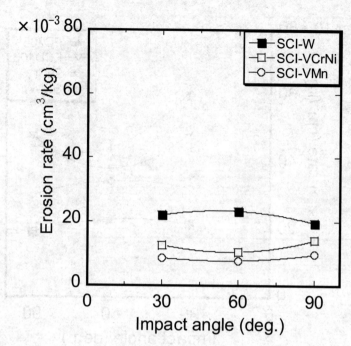

Fig. 3-3-1　Erosion rate vs. Impact angle in three kinds of spheroidal carbides cast irons

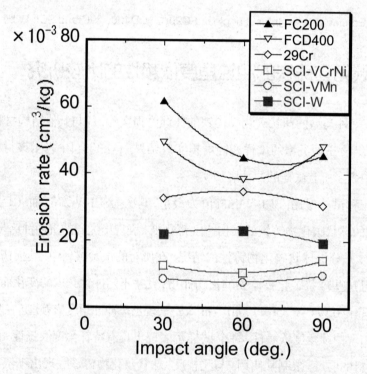

Fig. 3-3-2　Erosion rate vs. Impact angle in all specimens

3.4　磨损试验后的硬度测定

为考察这些结果与磨损表面硬度变化的关联性,利用维氏硬度试验测试试验前后的硬度变化。硬度测试结果表示于Fig.3-4-1中。试验材料的初始硬度在Table3-2-1和Table3-2-2中表示。FC200和FCD400的初始硬度低,由测定加工硬化后的表面硬度得知,硬度小于300 HV,这比冲蚀颗粒石英砂的硬度低很多,所以被认为不能抵抗冲蚀颗粒的连续碰撞。相比之下，29Cr显示出较高的初始硬度值（642 HV），但在试验后硬度不仅没有增加，其损伤速度也比SCI系列铸铁多。所以，在三种类型的SCI系列铸铁中，经过硬度测试，得知所有材料发生了显著的加工硬化现象。

在SCI-VCrNi中，试验前的维氏硬度是399 HV，试验后测得硬度为482 HV，约增加了20%，可以理解发生了明显的加工硬化。

在损伤速度值为最小的SCI-VMn中，看出试验后的硬度804 HV比试验前的硬度504 HV约增加了60%。为什么SCI-VMn的硬度显著提高，其原因可以有如下解释：其基体是奥氏体，且含有大量的Mn元素，在冲蚀磨损进行过程中，受到冲蚀颗粒的碰撞时，奥氏体产生应变诱导相变后转变成马氏体，导致表面硬度显著提高，从而贡献与抗冲蚀磨损性的提高。

在SCI-W中也可看到，试验后的硬度比试验前的硬度提高了不少。其表面硬度在3 600 s试验后，硬度从初始硬度418 HV提高到试验后的607 HV，约增加了45%。

因此，冲蚀磨损是依赖于材料的硬度不假，但是更重要的可以理解为，与其说依赖于初始硬度，还不如说更依赖于加工硬化后的硬度。

冲蚀颗粒的硬度和形状等外界因素会影响材料的损伤速度，但就开发抗冲蚀磨损材料而言，得到硬质的碳化物均匀分布在基体中，所以用基体发生形变诱导相变来提高材料的硬度也是众望所归。

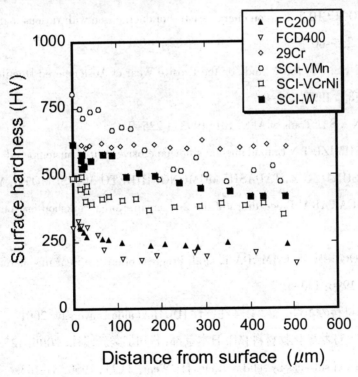

Fig. 3-4-1　Change of Vickers hardness for specimens after erosion test

3.5　小结

在这一章中，进行了三种类型的球状碳化物铸铁的冲蚀磨损试验。其结果，在球状碳化物铸铁中，由于球形VC的均匀分布, 磨耗损失显著减少。更重要的是, 所有材料都没有表现出冲蚀角度依赖性, 且具有良好的抗冲蚀磨损性能。这一结果, 相对于先前的钢铁材料或铸铁材料存在着鲜明的冲蚀角度依赖性特征来说, 是崭新的, 目前也没有发现相关的类似的研究报道。

冲蚀颗粒用钢砂和石英砂, 对FC200, FCD400, 29Cr, SCI–VCrNi, SCI–VMn和SCI–W 等六种实验材料进行冲蚀磨损性能评价, 其结果如下:

(1) SCI–VCrNi比高铬铸铁显著提高了冲蚀耐磨性能。

(2) SCI–VCrNi的磨损表面维氏硬度 (482 HV) 比初始硬度 (399 HV) 更硬。可以确定, 由于颗粒的碰撞, 发生加工硬化显著提高了表面硬度, 从而改善其抗冲蚀耐磨性。

(3) SCI–VMn的磨损量是磨损量最大的FC200的1/8, 显示出良好的抗冲蚀磨损性能。

(4) 按升序排列, 抗冲蚀磨损性能为: SCI–VMn > SCI–VCrNi > SCI–W > 29Cr > FCD400 > FC200。

(5) 冲蚀磨损中与其说初始硬度重要, 还不如说加工硬化后的硬度更重要。越容易发生形变诱导相变来提高硬度的材料, 越表现出优异的抗冲蚀耐磨损特性。

参考文献

[1] SHIMIZU K, NOGUCHI T. Erosion characteristics of ductile iron with various matrix structures [J]. Wear, 176 (2), 1994: 255–60.

[2] K.SHIMIZU T N, S.DOI. Basic Study on the Erosive Wear of Austempered Ductile Iron [J]. Transactions of AFS, 101 (93–78), 1993: 225–9.

[3] K.SHIMIZU T N A S D. Trans of AFS, 101 (1993): 225–9.

[4] KAZUMICHI SHIMIZU T N E. fundamental study on erosive wear of austempered ductile iron [J].

[5] KAZUMICHI SHIMIZU X, TADASHI MOMONO, HIDETO MATSUMOTO, YOSHIYUKI MAEDA, KIYOSUKE SUGAWARA. Proceedings of the 2nd Japan–Korea workshop for young foundry engineers [J]. 2005: 1.

[6] SHIMIZU K, NOGUCHI T, KAMADA T, et al. Progress of erosive wear in spheroidal graphite cast iron [J]. Wear, 198 (1), 1996: 150–5.

[7] 清水一道. 鉄鋼材料のエロージョン摩耗特性 [D]. Hokkaido University, 2001.

[8] 球相材料研究会. 球状碳化物材料 [M]. 日本京都: 日刊工業新聞社, 2006: 125.

[9] FINNIE I. Erosion of surfaces by solid particles [J]. Wear, 3 (2), 1960: 87–103.

第4章 球状碳化物铸铁的冲蚀磨损机理

4.1 序言

在第3章中, 使用抽吸式喷砂机, 以平均粒径为408 μm 的石英砂作为冲蚀颗粒, 进行了冲蚀磨损实验, 阐明了冲蚀角度和冲蚀速度的关系。随着冲蚀角度的变化, 各种材料表现出不同的冲蚀速度, 即不同的冲蚀角度依赖性。但是, 开发的几种球状碳化物铸铁的共同点, 即已发现该材料的冲蚀角度依赖性很小或几乎没有。因此, 为了阐明球状碳化物铸铁的冲蚀磨损机理, 本章利用金相显微镜直接观察磨损表面和磨损横截面连续照片, 并在扫描电子显微镜下观察磨损表面。此外, 利用双试样截面连续观察法, 观察切成两半的试样重合后进行每个时间段的磨损试验, 并观察每个预定时间段的横截面金相照片, 从而试图理解和把握详细的冲蚀磨损过程。

4.2 试样准备及试验方法

使用的试样有, FC200, FCD400, SCI–VCrNi, SCI–VMn和SCI–W, 都是在第3章的冲蚀实验中所使用的50 mm×50 mm×10 mm的试样。首先, 为了宏观观察磨损表面, 把3 600 s冲蚀磨损时间后的试样片的磨损表面用数码相机进行拍照。然后, 微观观察磨损表面照片(扫描电子显微镜照片: Scanning Electron Microscope)。最后, 为了观察磨损表面附近的组织变化, 进行了横截面连续观察。此外, 测量磨损表面的维氏硬度, 一起讨论。

4.3 冲蚀磨损表面宏观观察

为了阐明各试样的磨损机理, 观察了冲蚀实验后的冲蚀磨损宏观表面。Fig.4-3-1中表示球形钢砂的情况, Fig.4-3-2和Fig.4-3-3表示不规则形状的硅砂的情况及冲蚀磨损表面的宏观照片。照片中表示冲蚀磨损时间3 600 s后的情况, 冲蚀角度为30°, 60°, 90° 的磨损表面。所有试样表面出现类似圆形或椭圆形的损伤痕迹区域, 这个区域被认为是大致相当于喷射粒子的碰撞材料表面的范围。根据清水一道的研究及其他研究人员的研究结果[1-14], 可以得知使用钢砂做冲蚀磨损实验时, 在靶材S45C磨损表面观察到了垂直于冲蚀方向的清晰的条形纹理。

然而, 在本研究中, 如Fig.4-3-1所示, 在利用球形钢砂的情况下, FC200和FCD400中没有观察到例如

S45C那样明显的条纹,而且像SCI系列材料,由于冲击过程中硬度提高,只能观察到少量的磨损痕迹。如Fig.4-3-2和Fig.4-3-3所示,在使用硅砂情况下的冲蚀磨损宏观表面中,可以看出对于每个冲蚀角度,试样厚度方向的凹痕比使用钢砂的情况要多得多,但是,对所有的冲蚀角度都没有出现像钢砂碰撞S45C材料那样的条纹。

同时,对SCI-VCrNi, SCI-VMn和SCI-W而言,观察到对每个冲蚀角度,在磨损表面的深度方向上看,损伤深度明显变小。从这个事实也可以理解为,SCI系列铸铁的磨损体积比其他材料的磨损体积要小,也从侧面反映其损伤速度比其他材料大大下降,可以看出其优异的耐磨损性。但是,从30°到90°冲蚀角度,宏观上观察不到明确不同的表面纹理。为了说明低角度侧和高角度侧的冲蚀磨损机理的异同,有必要进一步从微观上观察其磨损表面情况。

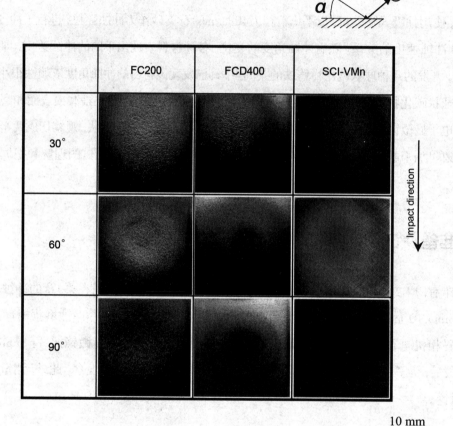

Fig. 4-3-1 Eroded surface in FC200, FCD400 and SCI-VMn by steel grits

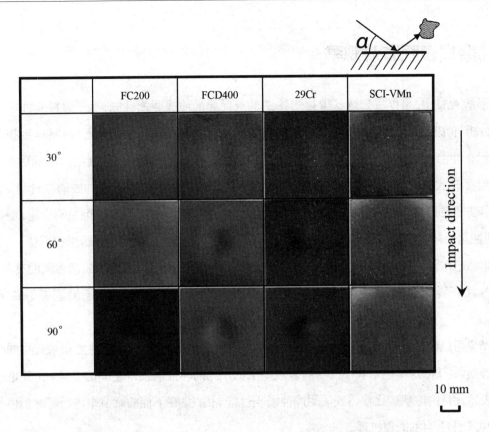

Fig. 4-3-2　Eroded surface of specimens by sand particles

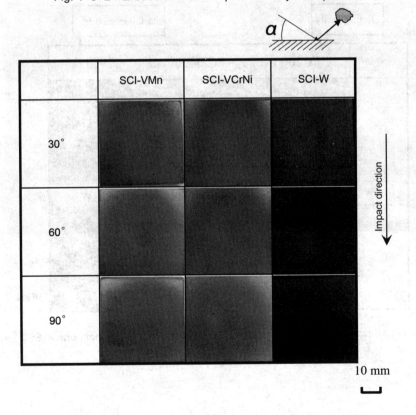

Fig. 4-3-3　Eroded surface of specimens by sand particles

4.4 冲蚀磨损表面微观观察

从宏观形貌观察中,可以看到球状碳化物铸铁具有优异的冲蚀磨损性能,为了解释其冲蚀磨损机理,需要进一步分析详细的微观磨损表面形貌。因此,我们通过SEM(扫描电子显微镜)观察微磨损表面形貌。Fig.4-4-1表示在冲蚀角度60°情况下的表面形貌。被磨损部位明显出现划痕,且划痕的末端形成舌状裂纹。与此同时,产生了较大的突起或凹陷。可以推断这种情况下的冲蚀磨损机理是,在冲蚀磨损过程中同时形成了变形磨损和切削磨损。Fig. 4-4-2和Fig.4-4-3磨损表面的SEM形貌图表示,随着冲蚀角的变化,磨损机理的改变规律。在冲蚀角度30°的情况下,在整个表面沿冲蚀方向上可以清楚地观察到被切割的痕迹。在另一方面,在90°冲蚀角度的情况下,可以看出磨损表面在垂直方向上塑性变形,磨损表面形成很多大的突起和凹陷。因此,可以断定,在倾斜角度的情况下,切削磨损占主导,随着冲蚀角度的升高,变形磨损逐渐变多,成为主要的磨损机理。

从上述结果可以得出结论,钢铁材料的冲蚀磨损的共性,即冲蚀磨损机理是在磨损过程中同时形成变形磨损和切割磨损。但是,这不能解释不同材料表现出来的不同的冲蚀磨损量之间的差异。尤其是也不能解释比一般铸铁表现优良的球状碳化物铸铁的耐冲蚀磨损性能。因此,在下面的章节中进行了磨损横截面的连续观察,试图把握靶材各自的磨损机理。

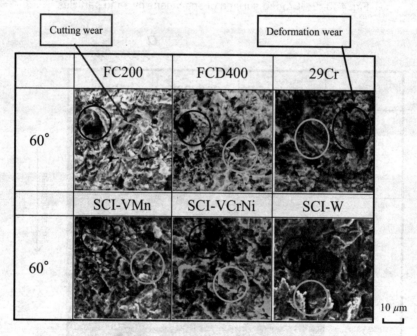

Fig.4-4-1   SEM observation on the eroded surfaces of specimens at 60°

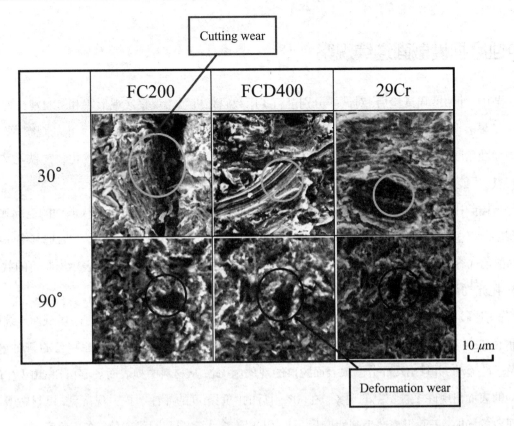

Fig. 4-4-2　SEM observation on the eroded surfaces of specimens at 30° and 90°

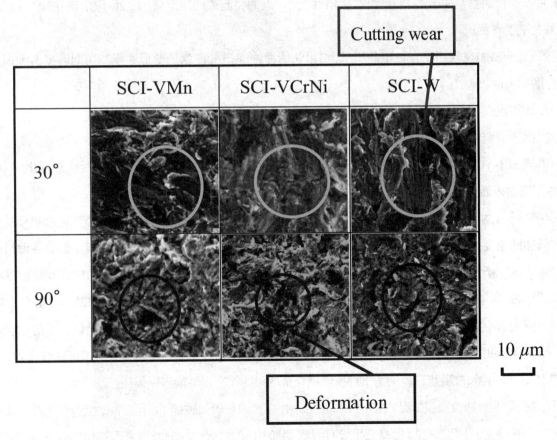

Fig.4-4-3　SEM observation on the eroded surfaces of specimens at 30° and 90°

4.5　冲蚀磨损横截面连续观察

在前一节中,冲蚀磨损试验后,对试样分别进行了用数码相机照相观察宏观形貌和SEM观察磨损表面的微观观察的结果表明,每种材料随冲蚀角度的不同具有不同的磨损机理。然而,迄今为止的观察中,无法解释靶材在不同冲蚀角度时出现不同的损伤速度的现象,尤其是本研究的重点——球状碳化物铸铁几乎没有角度依赖性的特点,不能得到满意的解释。

因此,在本节中,对已经利用SEM观察过的试样,进一步进行了磨损表面附近的横截面的连续观察。在横截面的观察中,首先利用精密切割机把试样沿着中心和试样被冲蚀的方向切成两半,然后把其中一半放置在事先预制的氯乙烯容器中央,并注入Araldite树脂来制备截面观察用样品。Fig.4-5-1表示出制备的截面样品。最后,研磨、抛光样品,用光学显微镜观察截面。

首先,作为比较材料,FC200,FCD400及29Cr的冲蚀角度分别为30°,60°,90°时的横截面连续照片的观察结果展示在Fig.4-5-2中。在所有样品中可以观察到,在30°的低角度侧,由于冲蚀颗粒的斜面冲击,材料的表面呈现塑性流动,并且出现部分磨损表面脱落的现象。由此,从材料磨损表面观察可知,由于冲蚀颗粒引起塑性流动的表面出现许多被切割的现象。在60°附近的中间角度的冲蚀磨损中观察到,以材料塑性流动而被切割的现象的同时,表面组织产生塑性变形,从而出现许多大突起和凹陷现象。在高角度侧90°冲蚀磨损时,由于冲蚀颗粒的垂直冲击,表面组织发生塑性变形,大的突起或凹陷形成。然而,像出现在30°的情况下切削现象几乎观察不到。

上述现象与SEM观察相结合讨论的结果,可以认为冲蚀磨损机理就是变形磨损和切削磨损两种损伤同时发生在磨损表面。

接着,详细观察、讨论球状碳化物铸铁的情况。

首先,描述研究SCI-VCrNi的情况如下:

为了阐明SCI-VCrNi的冲蚀磨损机理,下面结合前一章的讨论结果,将讨论磨损横截面状态变化。因此,同样观察了磨损表面的横截面形态。

Fig.4-5-3表示冲蚀角度分别为在30°,60°,90°时的SCI-VCrNi的磨损表面附近的横截面连续观察结果。首先,在低冲蚀角度30°的情况下,观察到该材料由于塑性流动,其表面沿着冲蚀方向流动,部分表面组织在表面脱落的同时,塑性流动的金属表面被切割的现象也不少。此外,还观察到在基体中的球状碳化物没有被切削掉而周围的基体组织有被切削的情况。由于硬质颗粒的撞击,观察到球状碳化物和基体一起脱落的现象。然而,塑性流动引起的向冲蚀方向突起的部位长度较短,突起的高度也比较低。可以确认坚硬的球状碳化物经受硬质颗粒的碰撞后也不变形,也不向冲蚀方向移动。由此可以判断,由于基体中有了球状碳化物,材料表面的基体结构的伸长被阻止,减小了表面脱落,该合金的耐冲蚀磨损性能得以提高。

接着,观察了90°冲蚀磨损情况。在90°的高角度侧,在冲蚀粒子的碰撞下,磨损表面组织产生塑性变形,形成了大面积的突起或凹陷现象,但是几乎没有观察到如30°的情况一样的磨损表面被切割的现象。虽然有大面积的突起或凹陷,但它是非常小的,突起长度也比其他材料小得多,所以观察到横截面是比较平滑的。此外,由于基体中的球状碳化物颗粒的垂直碰撞,出现裂开或脱落两种情况。

还有,在60°冲蚀角度的情况下,即损伤速度较低的情况下,在观察到由于塑性流动材料被切割的同时,表面组织塑性变形,从而形成突起和凹陷的现象。突起的长度比90°时的情况稍长一些。60°的情况属于切削和变形混合的磨损模式,和30°一样,可以观察到球状碳化物没有被切削掉而周围的基体组织被切削的情况。

从Fig.4-5-2中展示的试样中,重新对FCD进行讨论,并和SCI-VCrNi的情况相结合进行讨论。由于FCD中球状石墨周围的铁素体很软,而且石墨和铁素体的紧密性差,石墨和铁素体同时脱落形成大面积的凹凸磨损表面。和SEM照片中观察到的一样,可以确认低角度时为切削磨损,高角度时为变形磨损。而SCI-VCrNi球状碳化物铸铁中出现相对平滑的磨损横截面。可以看出,越是磨损量少的材料,越是凹凸小,磨损显著减少,从而磨损恶化被抑制。根据这一结果,SCI-VCrNi的磨损机理可以推断为,是变形的磨损和切削磨损两种机理同时作用的结果。然而,由于机体中有球状碳化物,使每个冲蚀角度的损伤速度得到了改善。原因在于,球状碳化物非常坚硬,加之受碰撞颗粒的撞击,材料表面的基体组织有了加工硬化,促使VC和基体之间接合得非常好。

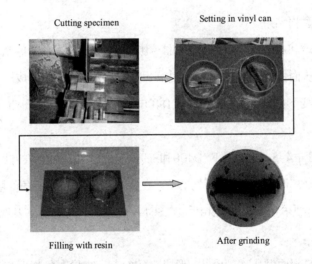

Fig.4-5-1　Procedure of making specimen for vertical section observation

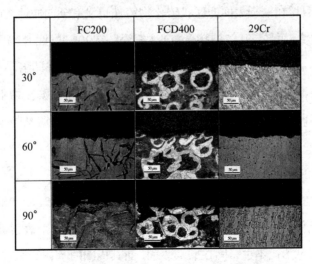

Fig.4-5-2　Vertical Section observation of FC200, FCD400 and 29Cr

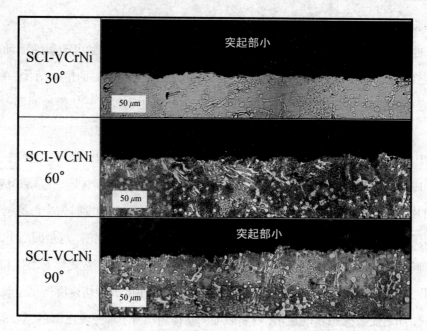

Fig.4–5–3   Vertical Section observation of SCI–VCrNi at 30° ,60° and 90°

接着，介绍SCI–VMn的磨损结果。

FCD400的组织中包括球状石墨，其周围有铁素体和珠光体。在基体中的球状石墨的硬度低，受到颗粒的冲蚀，容易发生塑性变形，从而使表面组织很容易脱落。相反，SCI–VMn中形成非常坚硬的球状碳化物，它具有分散应力集中、抑制塑性变形的效果。因此，以FCD400为比较材料，进行了SCI–VMn的截面显微观察，来讨论其冲蚀磨损机理。

Fig.4–5–4，Fig.4–5–5和Fig.4–5–6中表示FCD400和SCI–VMn使用硅砂作为冲蚀颗粒，冲蚀速度为100 m/s，冲蚀角度分别为30°，60°，90° 时的横截面照片。首先，在FCD400中，在所有角度磨损表面附近的石墨变形变成扁平状，出现清晰的塑性变形的痕迹。与此相反，在SCI–VMn中，不同角度在表现出不同磨损机理的同时，球状碳化物起到重要的作用。

在30° 的低角度时，材料表面发生切削磨损，形成小的突起。球状碳化物保持球形被留在磨损表面上，只有周围的基体被切割掉，并且留在磨损表面的球状碳化物铸铁也没有朝向冲蚀方向移动或流动。由于这些原因，可以理解为球状VC非常硬，而且与基体的接合力非常牢固。在90° 冲蚀的情况下，由于冲蚀颗粒的垂直碰撞，材料表面只有压缩力，球状碳化物和基体一起被挖掘，形成小的突起和凹陷。

此外，虽然球状碳化物断裂和脱落的两种情况都被确认，但是球状碳化物不发生塑性变形。随着基体的硬化，磨损的发展得到抑制。在60° 的冲蚀情况下，同时形成如30° 的切割磨损和90° 的脱落和开裂的磨损现象。

对于SCI–W的分析结果说明如下：

Fig.4–5–7中表示SCI–W的3 600 s冲蚀后横截面附近的组织变化。对于SCI–W的情况，观察的结果与之前其他球状碳化物铸铁非常相似。首先，在30° 低角度侧观察到该材料的整个表面被切削，形成了小的突起，同时也证实了球状碳化物保持球形比不过紧固在磨损表面，而其周围的基体组织被切削掉的现象。此外，由于SCI–W中的球状碳化物在铸造时出现了偏析，形成一些非常小的颗粒，可以认为这些小颗粒在冲蚀磨损过程中

容易与基体一起脱落,造成磨损表面形成凹凸不均匀。

在60°的情况下,形成的突起长度整体上小于30°的情况,还可以看到球状碳化物破裂后脱落的痕迹。在90°的情况下,因为材料表面仅受压缩力的作用,球状碳化物与基体组织同时被挖掘,受到冲蚀颗粒的碰撞后形成小的突起和凹陷。此外,还可以观察到球状碳化物的脱落和破裂的两种情况,球状碳化物没有发生塑性变形与基体的加工硬化,从而抑制磨损的发展。

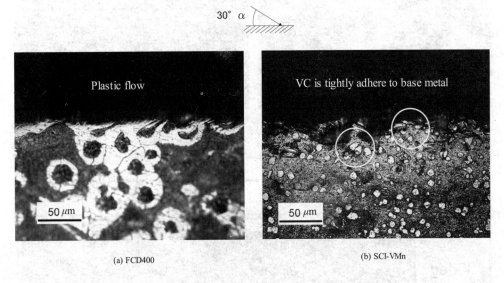

(a) FCD400　　(b) SCI-VMn

Fig.4-5-4　Cross section of FCD400 and SCI-VMn after erosion test at impact angles of 30°

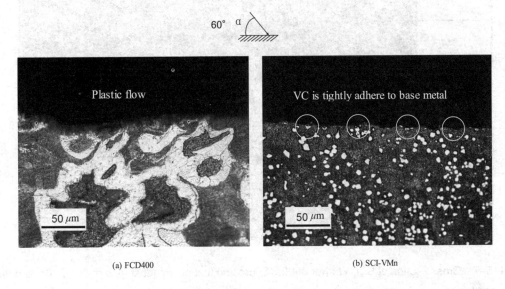

(a) FCD400　　(b) SCI-VMn

Fig.4-5-5　Cross section of FCD400 and SCI-VMn after erosion test at impact angles of 60°

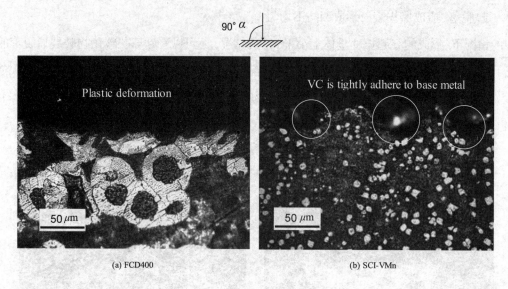

Fig.4-5-6 Cross section of FCD400 and SCI-VMn after erosion test at impact angles of 90°

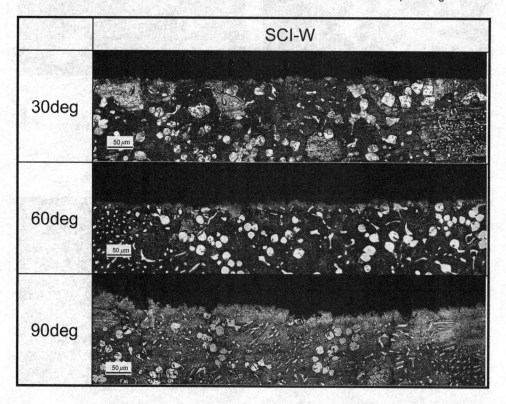

Fig.4-5-7 Cross section of SCI-W after 3600s. of erosion test at impact angles of 30°, 60° and 90°

4.6 两个半个试样的冲蚀磨损连续观察

在本节中,为了估算冲蚀磨损的寿命,在基体组织中,在含有球状VC碳化物的球状碳化物铸铁的磨损表面附近,按照不同的磨损时间连续观察其金相组织深度方向的变形、突起的形成、脱落方式等。连续观察方法如下:将两个半个试样的断面研磨抛光后,把断面合在一起放置,进行冲蚀磨损实验,连续观察磨损进程,同时计算出冲蚀磨损进程速度,以此来试图评估磨损寿命,最后验证其有效性。

因此,选用三种类型的球状碳化物铸铁,详细观察和讨论了它们冲蚀磨损进程,求出冲蚀磨损的进展速度,预测磨损寿命。

所用的实验材料为第2章中使用的球墨铸铁FCD400、高铬铸铁29Cr和第3章中的SCI-VCrNi和SCI-VMn两种球状碳化物铸铁。之所以使用FCD400,是因为基体组织中含有球状石墨,可以和本研究的球状碳化物铸铁进行比较。FCD400的球状石墨硬度很低,虽然其周围的珠光体硬度稍硬一些,但因为两者都在400 HV以下,在冲蚀粒子的碰撞下表面组织很容易变形,形成凸起和脱落。另一方面,SCI系铸铁中的VC是非常硬的,能达到2300 HV左右,而且它与周围的基体组织之间的接合力也非常好。

高铬铸铁中含有的碳化铬是非常硬的,但是形状以板条状或棱形为主,在冲蚀颗粒冲击下容易破裂且磨损容易进行。

所使用的实验装置和夹具如下:使用第2章中使用过的吸引式喷砂机为冲蚀磨损试验机,故不进行重复叙述,其详细结构见第2章中的关于试验机的描述。本次试样中,为了连续观察磨损表面附近深度方向的组织变化、突起部的形成、脱落状态,使用如下的小型特殊装置。Fig.4-6-1中表示其示意图,Fig.4-6-2中表示该装置的照片。实验用试样的形状和之前的常规冲蚀磨损实验中使用的试样相同,实验前将试样沿着其中心线切割成两半,所切割表面(试样的横截面)用金刚砂研磨、抛光,然后用4%的硝酸乙醇腐蚀液进行腐蚀,最后达到能在显微镜下观察为止。制备好的两半试样重新合并在一起,装入上述装置中,安装在试验台上准备实验。详细制备安装过程见Fig.4-6-3。冲蚀磨损方法与上一章的方法一致,只是到达一定时间的冲蚀磨损后,取出试样在金相显微镜下观察,如此隔一段时间观察一次,连续进行指定的次数。

根据第3章中Fig.3-3-2所示的实验结果,SCI-VMn和SCI-VCrNi的损伤速度为FCD400和29Cr的损伤速度的1/6~1/4。同时,也没有出现明显的冲蚀角度依赖性。很明显,球状碳化物铸铁表现出良好的冲蚀耐磨性能。可以看出,FCD400中低角度30°和高角度90°的磨损量都很高,表现出较弱的冲蚀磨损性能。在高铬铸铁中,虽然30°和90°的磨损量减少,但是在60°的情况下,表现出较高的磨损量值。与此相反,在SCI系列铸铁的冲蚀磨损中,可以看出30°~90°之间的所有角度,磨损量都得到了抑制。因此,连续观察损伤速度的变化,可以发现在变化比较大的30°和90°的情况下,磨损表面附近的深度方向上组织变化、突起形成、脱落状态,以此来详细研究其磨损机理。

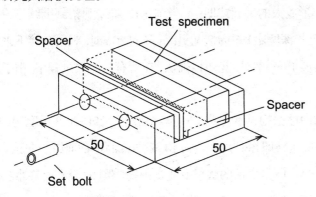

Fig.4-6-1　Sketch of the specimen stage for cross section observation (from ref.5)

Fig.4-6-2　The real specimen stage

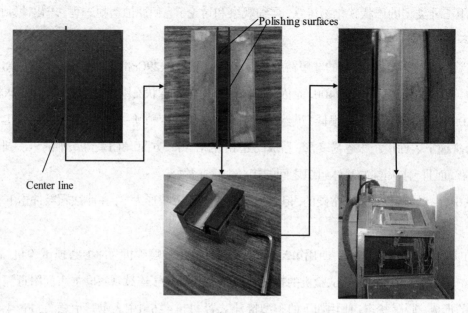

Center line

Polishing surfaces

Fig.4-6-3 Procedure in details of making two-sides pieces and its erosion test

接下来,详细描述各种材料在一定时间间隔内的磨损形态变化。

首先,使用FCD400进行了两个半个试样的冲蚀磨损实验。Fig.4-6-4表示在30°的低角度情况下的横截面组织连续变化的观察结果,其中冲蚀磨损时间分别为0 s, 5 s, 10 s和20 s的横截面组织变化结果。由于受冲蚀颗粒的冲击,虽然石墨和周围的基体沿着碰撞方向变形,但是这种变形比起使用钢砂的情况(根据前述Shimizu等研究)而言,向碰撞方向的突起较小。在5 s时,由于受到不规则形状的石英砂的碰撞,虽然开始形成突起部,但很快被切割而形成大的凹凸。当到达10 s时,由于碰撞继续,观察到在表面附近形成的凹凸不平,在深度方向加深的同时,出现突起被切削脱离表面的现象。到达20 s时,切削进一步加深,已经形成凹凸不平的一部分被切削脱落掉,并重新形成新的凹凸或凹陷。同时也观察到,如同在10 s时一样,球状石墨与基质一起被切割掉,形成凹陷变得更深的现象。

Fig.4-6-5表示在高角度90°的横截面组织连续变化的观察结果。与30°低角度一样,喷砂时间为0 s, 5 s, 10 s和20 s的横截面组织变化。在90°的情况下,由于冲蚀颗粒垂直方向撞击材料表面,在实验开始的5 s后,石墨与基体组织一起塑性变形,被压缩而破裂,形成大的突起或凹陷。当到达10 s时,观察到继续形成凹凸不平,已形成凹凸的长度比在30°时的情况下短,但其深度比30°的情况更深。超过20 s时,观察到接下来的冲蚀颗粒的碰撞,使已形成的突起继续破裂,磨损发展下去。另外,还可以确认,在所有的磨损过程中,由于压缩力的作用,球状石墨变形成为椭圆形。

下面描述关于高铬铸铁的两个半个试样的冲蚀磨损试验结果。Fig.4-6-6表示29Cr在低角度30°的情况下的横截面连续观察图。实验步骤类似于FCD400,喷砂时间为0 s, 5 s, 10 s和20 s的横截面组织变化。对于高铬铸铁,在0~5 s之间,虽然初始磨损的磨损量较大,但是之后的磨损变化量平稳。然而,没有向着冲蚀方向上呈现出由冲蚀颗粒的碰撞引起的塑性流动,而观察到由于石英砂的冲击而表面破裂现象。在5 s后,表面出现凹凸不平,沿着碰撞方向延伸,但是,长度比FCD400要短。随着冲蚀时间的延长,达到10 s, 20 s时,已经变成粗糙的表面,由于接下来受冲蚀颗粒的碰撞而大部分剥离、脱落,同时也形成新的突起,从而磨损进展。

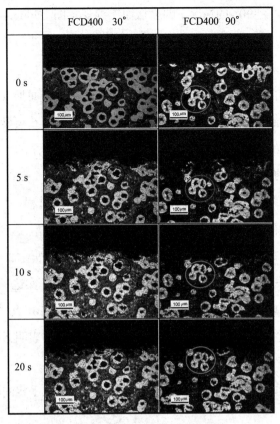

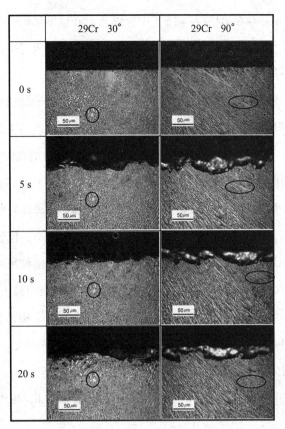

Fig.4-6-4　Microscopic continuous observation of vertical section in FCD400 at 30° and 90°

Fig.4-6-5　Microscopic continuous observation of vertical section in high chromium cast iron at 30° and 90°

　　Fig.4-6-7表示了29Cr的90°高角度的情况下横截面连续观察图。从图中可以观察到砂子碰撞材料表面使其压碎，但组织没有明显的塑性变形，从而形成大的突起。同时出现表面破裂，最后脱落掉等磨损过程。在从5 s到20 s的时间里，可以确认，从垂直方向以高速撞击的粒子，首先形成突起，然后通过下一个碰撞颗粒继续破裂，这样的过程重复进行来使材料表面磨损进行下去。还有，所形成的突出部的长度比30°的情况要短。虽然高铬铸铁的硬度远远超过了FCD400，但有一定的断裂韧性，所以，可以认为即使在90°的情况下也能抑制磨损的发展。

　　下面连续观察球状碳化物铸铁的组织变形、突出部形成及表面脱落形态。

　　对于SCI-VCrNi，Fig.4-6-6和Fig.4-6-7中分别表示30°和90°的情况下的横截面连续观察图。在Fig.4-6-6中，5 s时，通过从倾斜方向冲击的颗粒看，表面组织连同球状碳化物一起从磨损表面脱落掉，形成突起。当到达10 s时，已形成的突起由于受下一个冲蚀颗粒的碰撞，沿着碰撞方向向前延长，但是突起部位正下面的球状碳化物没有变形，而且沿着碰撞方向突出的突起部的长度也很短，没有观察到像FCD400中所观察到的那样清晰的伸长。同时，确认了新的突起的形成。时间到20 s时，有突起的形成和脱落继续发展，周围的基体被切削掉后球状碳化物出现在磨损表面的现象。所以可以认为，由于上述过程的重复进行使表面磨损。

　　再看90°的情况，如Fig.4-6-7所示，虽然0~5 s之间初始磨损较多，之后随着冲蚀时间的推移表层组织变形和脱落趋于平稳。在5 s时，表面层的组织变形，球状碳化物连同基体一起脱落，突起和压痕开始形成，到了10 s，突起进一步形成，并不断脱落。时间到达20 s时，突起越来越多，磨损继续发展。

Fig.4-6-8和Fig.4-6-9中分别表示在SCI-VMn中30°和90°的情况下的横截面连续观察图。如Fig.4-6-8所示,在30°的情况下,可以观察到由于初始磨损表面被切削,形成突起,表面上的球状碳化物将要与基体一起脱落的现象。

然而,从碰撞时间10~20 s的变化来看,由于球状碳化物的存在,受基体组织冲蚀颗粒的冲击而形成突出部,但是其朝向碰撞方向的伸长遇到球状碳化物后被迫停止。还有,由于球状碳化物非常硬,受到石英砂的冲击也不变形,只有其周围的基体出现被切削的结果时,在碰撞方向上的伸长长度缩短,受到接下来的冲蚀颗粒的冲击也很好地抑制了磨损,这个现象可以从20 s时的磨损情况中得到判断。

如Fig.4-6-9所示,在90°的情况下,观察到表面层的组织随着冲蚀时间逐渐地变形脱落的过程。在冲蚀5 s后,表面层的组织变形,球状碳化物连同基体一起脱落,形成突起和凹陷。到10 s后,进一步脱落。达到20 s时,突起几乎被脱落掉。从0~20 s的时间内,表面硬度从350 HV增加到600 HV。这种增加不算剧烈增加,可以认为是由于受冲击引起发生应变诱导硬化的表面被下一个粒子的冲击下部分脱落造成的。为了证实这一点,进一步观察横截面的结果,表面附近的横截面上确认出现马氏体的基体。其观察照片见Fig.4-6-10。此外,如Fig.4-6-11所示,试验前后磨损表面的XRD结果显示,磨损表面的γ-Fe比实验前减少了40%左右,表明磨损表面发生形变诱发相变,形成马氏体。

当比较低角度30°和高角度90°的磨损形式,也可以观察到都形成突起,突起部的伸长也几乎相同的结果。

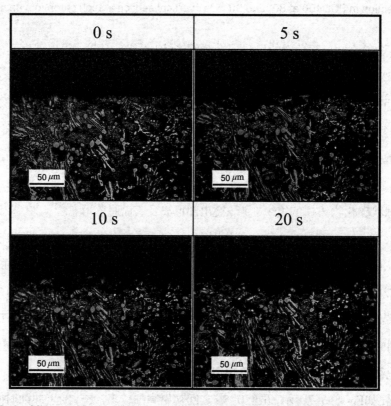

Fig.4-6-6　Microscopic continuous observation of vertical section in SCI-VCrNi at 30°

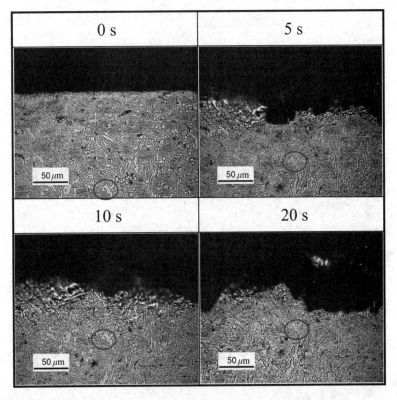

Fig.4-6-7 Microscopic continuous observation of vertical section in SCI–VCrNi at 90°

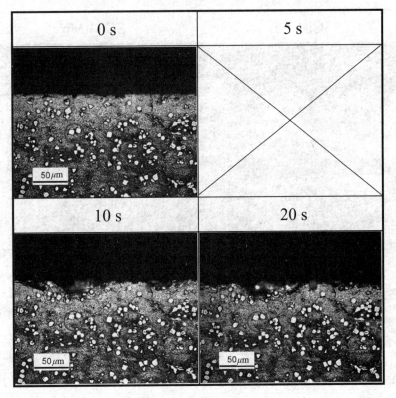

Fig.4-6-8 Microscopic continuous observation of vertical section in SCI–VMn at 30°

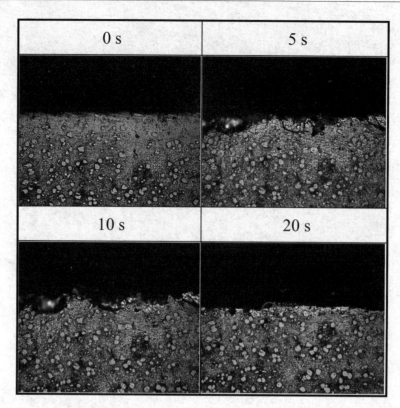

Fig.4-6-9　Microscopic continuous observation of vertical section in SCI-VMn at 90°

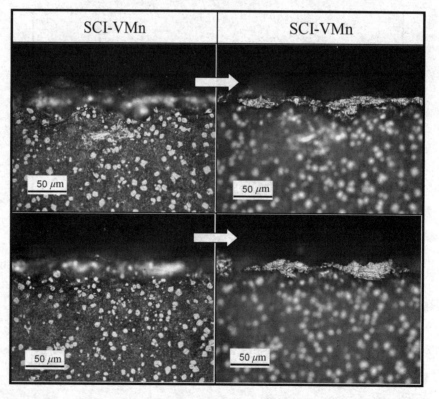

Fig.4-6-10　Martensite formed on the eroded surface

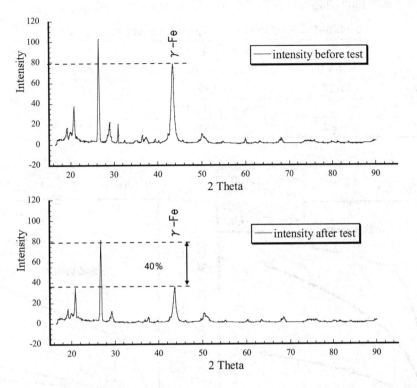

Fig.4-6-11　X-ray diffraction spectrum of SCI-VMn before and after erosion test

基于上述讨论,求出冲蚀磨损的进行速度。

根据清水一道等人的研究,结合本研究的讨论内容,测量出从磨损表面下的特定层到未磨损表面之间距离的变化,求出被去除的表面层的深度,也就是磨损深度,然后把它和磨损时间关联起来定义出磨损进展速度,从而用来估计实际的磨损寿命。在本研究中,根据该定义,求出每个材料的磨损进展速率,从而试图估算出材料的磨损寿命。

Fig.4-6-12表示关于SCI-VCrNi和SCI-VMn在冲蚀角度30°时的磨损深度和磨损时间的关系。冲蚀时间5 s以内属于初始磨损,之后随时间增加磨损深度按线性增加。SCI-VMn的磨损进展速率约为0.27 μm/s,SCI-VCrNi的磨损进展速率约为1.19 μm/s。

Fig.4-6-13表示关于SCI-VCrNi和SCI-VMn在冲蚀角度90°时的磨损深度和磨损时间的关系。初始磨损阶段为0~5 s,表面磨损深度迅速变小,之后则变得基本上恒定。SCI-VCrNi在0~5 s,磨损深度为迅速增至140 μm,表现为初始磨损大,之后线性增加。而SCI-VMn在 0~5 s,磨损深度只有7 μm,然后线性增加。该恒定增加的直线的斜率定义为磨损进展速率(μm/s)。磨损进展速率的值表示在一定量冲蚀颗粒下的材料厚度的减少速率。SCI-VMn的磨损进展速率约为1.06 μm/s,而SCI-VCrNi的磨损进展速率约为1.67 μm/s。因此,可以认为,根据某材料的磨损进展速率来估算该材料的实际磨损寿命。

在Fig.4-6-12和Fig.4-6-13中,表示包括其他比较材料FCD400和29Cr的所有实验材料的磨损进展速率。Table 4-6-1同时列出所有材料的磨损进展速率。从表格可以看出,SCI-VMn的磨损进展速率最为缓慢,它的值约为FCD400的1/6、高铬铸铁的1/4。这个结果与第3章中所得到的损伤速度的值相吻合。因此从磨损进展速率可以估算材料的实际磨损寿命。此外,更重要的是,SCI系列铸铁在磨损深度方向表现出具有优良的抗磨性能,这对实际工作中了解零部件的磨损具有非常重要的意义。

Table4-6-1　Comparison of progressing speed of erosive wear

	Impact angle of 30°			
Specimens	SCI-VMn	SCI-VCirNi	29Cr	FCD400
摩耗進展速度(μm/s)	0.27	1.19	1.65	3.44
	Impact angle of 90°			
摩耗進展速度(μm/s)	1.06	1.67	2.21	5.12

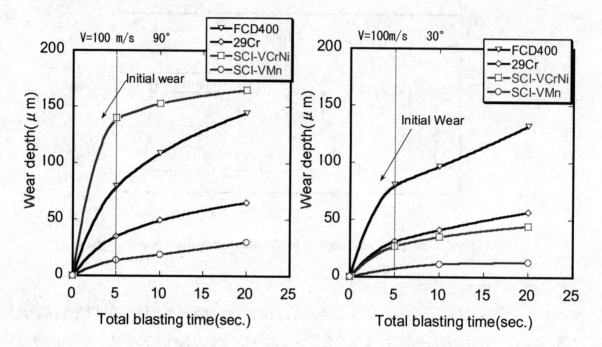

Fig.4-6-12　Wear depth vs. Total blasting time in specimen at impact angle of 30°

Fig.4-6-13　Wear depth vs. Total blasting time in specimen at impact angle of 90°

4.7　冲蚀磨损机理的模型建立

为了阐明球状碳化物铸铁的冲蚀磨损机理,在前几节中,进行了磨损表面的宏观、微观观察及冲蚀磨损表面附近的横截面观察。结果表明,球状碳化物铸铁和一般常见的结构材料,如球墨铸铁、高铬铸铁一样都具有共同点,那就是由于冲蚀颗粒的碰撞,材料表面形成突起,突起的伸长、脱落来磨损进展。但是,控制冲蚀磨损进展的重要原因是,球状碳化物铸铁中存在高硬度、球状的碳化物。本节中总结以上结果,试图建立冲蚀磨损机理的模型。

冲蚀磨损机理,如同从上述的SEM和横截面照片中可以看出的那样,可用切割磨损与变形磨损的和来表示。磨损进展取决于粉末和材料(特别是硬度)的事实越来越明显。在下文中,分别列出钢砂和石英砂作为冲蚀颗粒的情况下的FCD400、高铬铸铁及球状碳化物铸铁的磨损机理的示意图。

Fig.4-7-1(A),(B)中表示FCD400的磨损机理示意图。当冲蚀颗粒为钢砂的情况时,因为FCD400的硬度(240 HV)比钢砂(440 HV)低,它表面组织流动的纹理比较清晰。首先,在冲蚀角度小于90°($\alpha<90°$)时,通过冲蚀颗粒的冲击,磨损表面附近的组织向着冲蚀的下游侧塑性变形。这样向冲蚀方向的下游侧形成突起部,

它受冲蚀颗粒的冲击将大幅度增长，被随后的冲蚀粒子的碰撞而破裂、分离并脱落。磨损是通过这样的重复过程而进行的。这些突起连续产生，从而形成了有序的条纹。在高角度（α≈90°），表面层承受颗粒从垂直方向的冲击力而产生被压缩的塑性变形。然后，通过材料表面被压缩产生塑性变形，形成毛发状的突起，最后剥离的磨损机理。然而，如果靶材的硬度超过钢砂的硬度时，即高铬铸铁（29Cr）的情况，无论是在高角度还是低角度，因为受颗粒冲击引起的材料的塑性变形很小，冲蚀磨损快速下降。Fig.4-7-1（C），（D）中表示出其磨损示意图，没有条纹状形貌的出现。

在使用石英砂情况下的示意图在Fig.4-7-2和Fig.4-7-3中列出。在使用石英砂颗粒的情况下，表现出与在使用钢砂的情况完全不同的磨损类型。

首先，如Fig.4-7-2中所示，当低硬度材料FCD400在低角度（α<90°）冲蚀磨损时，最初受石英砂的冲击，材料表面发生塑性变形，朝向冲蚀方向形成向前延伸的突起。但是，由于这些突起的硬度很低，经下一个石英砂的冲蚀后，被切削，容易剥离掉，所以不出现像冲蚀颗粒为钢砂那样的清晰的条纹。因此，在低角度冲蚀时，以切削磨损为主流磨损方式。在高角度（α≈90°），由于石英砂是从垂直方向上冲击靶材，表面形成大的塑性变形和小的突出部。这部分经下一个颗粒冲击时，在达到其断裂韧性之前继续进一步塑性变形，直到从表面脱落，这样进行磨损。

对于高硬度材料，例如高铬铸铁（29Cr）的情况，在低角度时，虽然在石英砂的切削作用下形成突起，但突出部比突出部FCD400中形成的突起部要小得多，而且易碎。当遭到接下来的碰撞时，容易破裂并脱落掉。

在高角度时，受到垂直方向的冲击，表面破裂，形成压痕或凹陷，下一个颗粒的碰撞使凹陷扩展，形成新的新凹陷或突起，磨损继续进行。这种重复周期是非常快的。

以上叙述了延性材料和脆性材料的典型的磨损进展过程。

现在，建立磨损模型图，力图解释低硬度的基体中分散高硬度球状碳化物的球状碳化物铸铁的磨损机理。

Fig.4-7-3中表示球状碳化物铸铁的情况。

由于球状碳化物的存在，当受到冲蚀颗粒的冲击时，虽然磨损表面的基体组织变形，形成突出部，但是所形成的突起部位朝向碰撞方向的延伸遇到球状碳化物而被迫停止伸长。还有，由于球状碳化物非常硬，即使受到石英砂冲击也不变形，只有其周围的基体组织逐渐被切削后，突起的延伸才变小，即使受到下一颗粒的冲击，突起的伸长和脱落也会被很好地抑制住。因此，从低角度到高角度的全部过程中都可以观察到上述冲蚀磨损的共同点。如把这些共同点考虑到冲蚀磨损机理当中解释磨损机理如下：基体中存在球形碳化物，球形具有各向同性。由于冲击引起的应力集中，向各方向等同分散，所以从低角度到高角度在材料表面形成的凹陷基本相同，从而可以估计由此引起的综合塑性应变量也是一样的。这个现象可以从下面的两个过程来理解。

（1）假如把球状碳化物换成硬质的、不规则形状的碳化铬，那么，因为碳化铬的形状纵横比不同，在30°和90°冲蚀材料表面时，基体组织受到的应力就不同，因此碳化铬的尖端等容易应力集中的部位应力很大，导致冲蚀磨损量随冲蚀角度不同而不同。再者，如果把碳化铬改成球状石墨，最初，因为球形石墨的存在，不管低角度还是高角度冲蚀时，可以认为从球状石墨施加给基体的应力是相等的，但是下一个颗粒碰撞时，软质的球状石墨开始变形，而且如同前一节中观察到的一样，低角度和高角度的变形是不一样的。所以认为不同形状的石墨作用于基体的应力也是不同的，最终和碳化铬一样，在不同的角度冲蚀时，呈现出不同的磨损量。

最后,回到球状碳化物,由于球形的碳化物硬度达到2300 HV,远远超过石英砂的硬度,受到冲蚀时不仅不变形,而且在不同角度下球状碳化物对基体组织施加的应力是相等的。因此,认为不管在30°冲蚀还是在90°冲蚀,都应该表现出相同的磨损形态。

(2)从基体组织的角度考虑,受到冲蚀颗粒的冲蚀时,基体组织产生加工硬化。FCD400的情况,虽然有加工硬化,但加工硬化后的硬度不超过400 HV,冲蚀磨损形态不变。对于含Cr合金来说,即使受到冲击,其硬度几乎不会改变,所以没有基体组织加工硬化的效果。另一方面,在SCI的情况下,其基体结构受到冲击后,出现显著的加工硬化现象,所以可以推断,由于基体组织的加工硬化,球状碳化物朝向冲蚀方向的移动困难,碳化物与基体组织之间的黏合力更加牢固,球状碳化物控制磨损的效果更加提高。

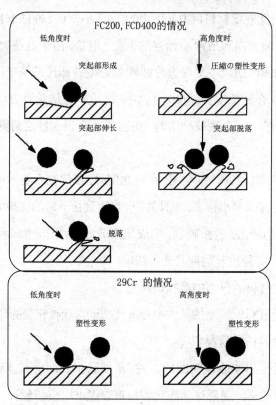

Fig. 4-7-1 Model for materials with steel grits
((A)(B)model of FCD400,(C)(D)model of Cr alloy)

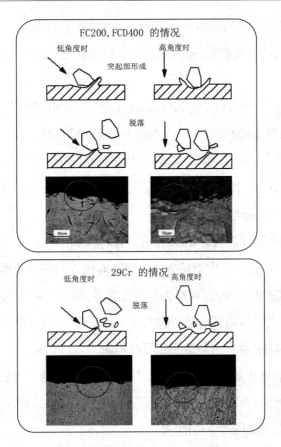

Fig.4-7-2　Model for materials with sand particles

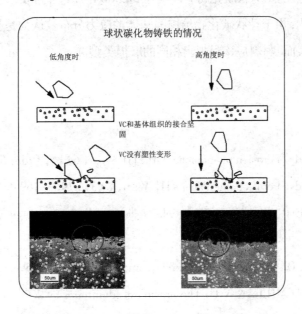

Fig.4-7-3　Model for SCI series with sand particles

4.8　小结

　　在本章中, 为了阐明各种球状碳化物铸铁的冲蚀磨损机理, 直接观察磨损表面, 表面附近横截面连续照片和磨损表面SEM照片。进一步, 利用两个半个试样的冲蚀磨损的金相显微镜分段连续观察, 理解冲蚀磨损进展的详细过程,最后,力图建立冲蚀磨损机理模型来解释冲蚀磨损机理。

下面总结冲蚀磨损试验后的磨损表面,根据表面附近横截面连续照片和磨损表面SEM照片观察得到结论。

在冲蚀颗粒为钢砂的情况下,靶材FC200和FCD400的磨损表面出现明显的条纹状形貌,冲击后硬度显著提高的SCI系列靶材,磨损表面上的磨痕很小,也观察不到条纹状形貌。与其他比较材料相比,SCI-VCrNi、SCI-VMn和SCI-W靶材中,对于每个冲蚀角度,在磨损表面的深度方向的损伤深度变浅,也可以看出随冲蚀角度的增加,磨损表面面积也变小,故可以理解为表现出优秀的冲蚀磨损性能。但是,当冲蚀角度为30°到90°时,从表面宏观形貌中看不出有相互间的不同之处。

其冲蚀磨损机理可以认为是切削磨损和变形磨损同时发生的结果。

此外,详细观察和讨论磨损进展,寻求能够应用于耐磨寿命评估的冲蚀磨损进展速度,并证实了它的有效性。其磨损机理是,由于充实颗粒的冲击,材料表面塑性变形、脱落、加工硬化、硬化层形成、再脱落等过程的重复呈现。根据磨损进展速度能够估算出管道的内壁厚度变薄,所以可以大致计算出其使用寿命。也就是说,利用磨损进展速率,能够简易地估计材料的实际磨损寿命。

在低冲蚀角度,由于加工硬化表面硬度得到提高,从而抑制磨损,这可能是主流抗磨形式。在高角度的情况下,受冲击的磨损表面加工硬化,而且在激烈的冲击下加工硬化层随后被破坏,但由于组织中含有球状碳化物,使应力集中分散,吸收能量,从而能够抑制磨损。

由于在SCI系列铸铁的组织中,球状碳化物的硬度达到2300 HV,它非常硬,即使受强烈冲击也不变形。还有,从低角度到高角度范围内,由于球状碳化物给周围基体的应力分布被认为是相同的,因此,冲蚀颗粒不管从30°还是90°冲击该材料表面,材料应该呈现出相同的磨损类型。

参考文献

[1] BITTER J G A. A study of erosion phenomena part I [J]. Wear, 6(1), 1963: 5–21.

[2] BITTER J G A. A study of erosion phenomena [J]. Wear, 6(3), 1963: 169–90.

[3] WINTER R E, HUTCHINGS I M. Solid particle erosion studies using single angular particles [J]. Wear, 29(2), 1974: 181–94.

[4] MISRA A, FINNIE I. On the size effect in abrasive and erosive wear [J]. Wear, 65(3), 1981: 359–73.

[5] RICKERBY D G, MACMILLAN N H. The erosion of aluminum by solid particle impingement at oblique incidence [J]. Wear, 79(2), 1982: 171–90.

[6] COUSENS A K, HUTCHINGS I M. A critical study of the erosion of an aluminium alloy by solid spherical particles at normal impingement [J]. Wear, 88(3), 1983: 335–48.

[7] SUNDARARAJAN G, SHEWMON P G. A new model for the erosion of metals at normal incidence [J]. Wear, 84(2), 1983: 237–58.

[8] 清水一道. 鉄鋼材料のエロージョン摩耗特性 [D]. Hokkaido University, 2001.

[9] K.SHIMIZU T N A S D. Trans of AFS, 101(1993): 225-9.

[10] KAZUMICHI SHIMIZU, NOGUCHI T. IMONO, 66(1994): 7.

[11] SHIMIZU K, NOGUCHI T. Erosion characteristics of ductile iron with various matrix structures [J]. Wear, 176(2), 1994: 255-60.

[12] SHIMIZU K, NOGUCHI T, KAMADA T, et al. Progress of erosive wear in spheroidal graphite cast iron [J]. Wear, 198(1), 1996: 150-5.

[13] KAZUMICHI SHIMIZU, NOGUCHI T. Formation and Progression of Sand-Erosion Surface in Spheroidal Ductile Iron [J]. TransJpnSocMechEng, 65(632), 1999: 940-5.

[14] WOOD R J K, WHEELER D W, LEJEAU D C, et al. Sand erosion performance of CVD boron carbide coated tungsten carbide [J]. Wear, 233-235(1999): 134-50.

第5章　新型Fe基Fe–V–Cr–Mn合金的磨粒磨损性能

5.1　磨损现象

5.1.1　磨损

磨损[1, 2]是一种复杂的现象,目前还没有对其精确和统一的定义。英国机械工程师协会定义为:磨损是由机械作用而造成物体表面的逐渐损耗。克拉盖尔斯基的定义为:由于摩擦结合力反复扰动而造成的材料的破坏。我国邵荷生则认为:由于机械作用、间或伴有化学或电的作用,物体工作表面材料在相对运动中不断损耗的现象称为磨损。[3]

磨损习惯于分为以下几个类型:

1. 磨粒磨损(磨料磨损)　由于硬质物料或突出物与表面相互摩擦使材料发生损耗的现象。

按照不同的分类方式,磨粒磨损又可以分成不同的磨损形式,其中按磨料对材料的力学作用特点可以把磨粒磨损分为以下3个方面:

(1)凿削磨粒磨损　在较严重冲击载荷下的磨粒磨损,例如颚式破碎机颚板上发生的磨损等。

(2)研磨磨粒磨损　在高应力下的磨粒磨损,例如球磨机磨球和衬板的磨损等。

(3)刮伤磨粒磨损　在低应力下的磨粒磨损,也称作冲蚀磨损或冲刷磨损,例如渣浆泵叶轮和护套的磨损等。

2. 腐蚀磨损　环境介质与材料表面发生的化学或电化学反应,伴随机械作用使材料损失的现象。

3. 黏着磨损　在黏着力的作用下使材料在两个表面发生迁移的现象。

4. 疲劳磨损　由于交变应力的不断作用使材料出现疲劳脱落的现象。

在实际生产中,磨损往往并不是以单一类型存在,而是几种磨损形式同时存在,相互影响,但是总有一种起主导作用。所以在分析材料磨损情况时,首先要弄清起主要作用的磨损类型和磨损机理,再结合材料的使用环境及使用条件,得出材料磨损的原因,并提出具体的改进方案。

5.1.2　磨粒磨损机理的研究

本节主要介绍磨粒磨损的磨损机理,主要有以下几种:

1. 微观切削机理[4]

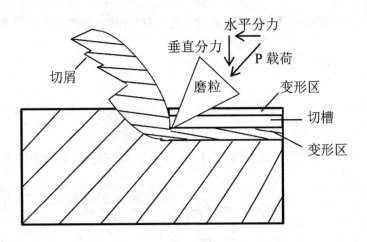

Fig. 5-1-1　Micro cutting model

将磨粒作用于材料表面的力分解为法向分力（正压力）和切向分力（摩擦力），如Fig.5-1-1所示。

在法向力的作用下，磨粒会被压入接触表面，形成压痕；在切向力的作用下，磨粒被向前推进。微观切削理论适用于作用在材料表面的磨粒具有锐利棱角和适当迎角的情况。当夹角太小或者材料塑性较高时，并不会削出磨屑，只会在材料表面犁出一条沟来。

2. 犁沟变形机理

在微观切削机理中提到的当磨粒圆钝与接触表面并没有合适的夹角时，磨损机理主要为犁沟变形机理，如Fig.5-1-2所示。

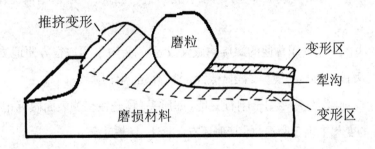

Fig. 5-1-2　Furrow deformation model

压入接触表面的磨粒在切向分力的作用下，会使材料向前或者向两侧堆积，并使这些区域产生塑性变形，后续的磨粒如此反复，可能把堆积起来的材料压平，也可能使材料再次发生犁沟变形，最终导致裂纹、断裂，表面形成磨屑后脱落。

3. 疲劳磨损机理

摩擦过程中接触峰在循环应力作用下，因疲劳剥落而形成磨屑。克拉盖尔斯基[5]认为疲劳磨损机理在一般磨粒磨损中起主导作用。但实验表明，疲劳极限不是对金属耐疲劳磨损性能的判据，关于疲劳破坏本质的经典概念不能用于磨料磨损机理。

4. 微观断裂（剥落）磨损机理

材料在这种情况下的磨损量通常较大，这是由于磨损过程往往伴随有塑性变形，但一些脆性材料，比如陶瓷、玻璃等，或者一些比较硬脆的第二相质点的材料与磨料作用时，会产生比较严重的断裂现象，或使材料直接剥落。

5.1.3 磨损量评定法

评定材料磨损的三个基本磨损量是长度磨损量（线磨损）、体积磨损量（体积磨损）和重量磨损量（质量磨损），见Table5-1-1。

Table5-1-1 Basic wear removal loss

名称	符号	单位	定义
线磨损	WL	mm	摩擦表面法线方向的尺寸损失
体积磨损	WV	mm^3	磨损表面的体积损失
质量磨损	WW	g或mg	磨损表面的质量损失

1. 失重法

通过称量试样磨损实验前后的质量变化来确定磨损量，一般在高精度天平上进行。由于操作简单，并且可以进行精度较高的测定，因此应用最为广泛。对密度不同的材料，应用磨损失重除以材料密度换算成磨损体积再加以比较更为合理。

2. 磨损尺寸变化测定法

对一些实际零件，则往往对某些接触部位尺寸变化情形进行测量。一般根据情况，可以采用宏观或微观测量方法来测量尺寸的变化。

（1）宏观尺寸测量法 一般选用普通的螺旋测微仪或者游标卡尺，可以很方便地测出材料磨损后的尺寸变化量。其他如测长仪、万能工具显微镜、读数显微镜、投影仪也可使用。

（2）微观测定法——刻痕法 用专用金刚石工具（如压头）预先在待测表面做标记（如压痕或磨痕），磨损后可以测量刻痕尺寸的变化来确定磨损量。其他还有台阶法、切槽法等。

（3）磨痕宽度测量法 在有些情况下，特别是在润滑条件下，磨损量很小，普通天平测不准重量变化，又不适宜用其他方法时，可以通过在显微镜下测量磨痕的宽度来评价磨损量的大小。

（4）微观测定法——表面形貌测定法 通过测量磨损表面粗糙度的变化来反应材料尺寸的变化，一般用到的仪器为轮廓仪。

3. 放射性同位素测定法

通过测量磨损材料及磨屑中单位时间内的原子衰变数，来反应材料的磨损变化。

4. 磨屑收集及测定法

磨损产物是材料磨损过程的最终结果，它综合反映了材料在磨损过程中的机械、物理、化学的作用和变化。它比磨损表面更直接地反映了磨损的机理和原因，因此，磨屑检测分析技术在磨损评定中显得格外重要。磨屑收集的方法很多，例如，将润滑油中所含的磨屑经过滤或沉淀分离出来，对所收集到的磨屑颗粒可以用各种方法进行定量分析。其他方法还有化学分析法、铁谱分析法、放射性同位素法等。

5.1.4　磨损率评定法

在所有情况下,磨损都是时间的函数,因此,有时也用磨损率来表示磨损的特性。例如,单位时间的磨损量、单位摩擦距离的磨损量、单位摩擦转数的磨损量及比磨损量(率)等,见Table5-1-2。

Table 5-1-2　Wear rate

磨损率	符号	单位	定义
单位时间的磨损量	Wrt	mg/s或mm^3/s	磨损量与摩擦时间之比
单位摩擦距离的磨损量	Wrs	mg/m	磨损量与摩擦距离之比
单位摩擦转数的磨损量	Wrr	mg/r	磨损量与摩擦转数之比
比磨损量	Wap	$mm^3/N \cdot mm$或$mm^3/kgf \cdot mm$	体积磨损与载荷、摩擦距离乘积之比

也可用耐磨性来辨别材料抗磨损性能的强弱。耐磨性可以分为两种:相对耐磨性和绝对耐磨性。

材料的相对耐磨性是指在相同的外部条件下,两种材料的磨损量比值,假设两种材料A和B,选材料A为参考试样,见公式(1-1),

$$E = W_A / W_B \tag{1-1}$$

其中,W_A和W_B一般代表的是体积磨损量,也可以表示为其他的磨损量。

绝对耐磨性通常用磨损量或磨损率的倒数来表示,见公式(1-2),

$$W^{-1} = 1/W \tag{1-2}$$

耐磨性使用最多的是体积磨损量的倒数。

5.1.5　不同评价方法的分析比较

用磨损量来评价磨损的方法,由于每个实验者所测试的试样形状、尺寸并不完全相同,且不能反映出磨损实验时的载荷、速度和摩擦距离,所以对其他研究者不具有较强的参考性,但对实验者自己而言,可以对实验材料进行相应的横向与纵向比较。

磨损率作为一个磨损量与一定磨损长度的比值,也无法直观地反映出实验所施加的载荷和速度。

相对耐磨性需要一个试样作为标准试样,但由于现在磨损试验机的种类比较繁多,所需的标准试样尺寸不一,摩擦副的表面粗糙度也不相同,不便要求实验者采用外形、尺寸和性能相同的试样作为标准试样。

目前国内外使用较普遍的是比磨损量来评价材料的磨损性能。这是因为比磨损量考虑了实验时所施加的载荷和摩擦距离的因素,可以在相同滑动速度的条件下进行比较。使用比磨损量来衡量材料的磨损性能还有一个好处,就是可以直观地通过它的数值大小来评价磨损的严重程度。日本学者认为,比磨损量在$10^{-8} mm^3/(N \cdot mm)$以下为轻微磨损,以上则为严重磨损。我国王铀[2]等人通过研究更进一步指出,比磨损量在$(1 \times 10^{-8}) \sim (2 \times 10^{-3}) mm^3/(N \cdot mm)$之间为以轻微磨损向严重磨损转化的界限,在这之间区域的数据被认为是过渡状态,称作过渡磨损。

5.2 新型Fe基Fe-V-Cr-Mn合金的开发设计

5.2.1 引言

本课题的目的是在球状碳化物铸铁的研究基础上[6],适当调整V, Cr, Mn等合金元素的成分配比,在不降低合金性能的前提下尽量减少贵重金属,尤其是钒的加入量。新合金的基体主要以奥氏体为主,在基体上形成弥散分布的高硬度质点颗粒,同时增强基体的形变硬化能力,最终得到一种硬度较高、韧性较好的耐磨新材质,使之与普通耐磨材料相比,耐磨性有2~4倍的提升,具有更好的机械性能。通过合金设计,铸造工艺方案设定及熔炼技术来制备高性能的金属耐磨材料。为了探讨球状碳化物的组成、构造,以及其形成条件,将对析出球状碳化物的结构、形状、大小、分布进行详细的调查研究。为了积极析出钒的球状碳化物的同时抑制其他种类碳化物的析出,将所添加的合金元素对形成球状碳化钒的影响进行分析,从而调整合金元素的加入量及成分。

由于工业的迅猛发展,球磨机被广泛用于化工、采矿、建筑等领域,用于粉磨各种矿石及其他物料。其中对球磨机衬板及磨球的需求量是最多的,同时对它们的磨损也是最大的。现有的衬板、磨球大部分选用高铬铸铁、高锰钢等材料,可以一定程度地满足球磨机的使用要求,但是大部分的使用寿命还是偏短,耐磨性偏低。有些耐磨材料还需要经过热处理后才能使用,增加了生产成本。因此,研制出一种新型合金耐磨材料,在铸态下就拥有比普通耐磨材料好1~2倍的耐磨性,同时具有较好的冲击韧性是非常有意义的。

5.2.2 合金设计与制备方案

5.2.2.1 合金的设计

合金元素在金属材料中发挥了极其重要的作用,它们既影响了材料的组织形态,又影响了材料的自身性能。以下将对本材质中用到的几种元素做简单的介绍。

碳 碳是决定材料硬度和韧性最为重要的元素。它能提高淬透性,影响显微组织。碳含量大小应根据使用条件不同而有所不同。在冲击载荷不是很大时,应选用中碳或高碳;而在高冲击载荷条件下,应选用低碳。另外,碳可以与材料中的一些合金元素,如钒、铬、钛、铌、锆等形成碳化物,使材料具有一定的特殊性能。提高合金碳含量,比增加合金含量更能增加组织中碳化物的数量。以铬为例,铬的碳化物体积分数K%可以用公式(2-1)估算:

$$K\%=11.3(C\%)+0.5(Cr\%)-13.4 \tag{2-1}$$

其中C%和Cr%为质量分数。随着碳含量提高,材料的硬度也提高,碳化物数量增加,抗磨性得到加强,但韧度降低。

硅 硅的固溶强化作用大于锰、铬、镍、钒,可以改善共晶碳化物的形态,提高Ms点,减少残余奥氏体。当含量超过3%时,将使材料的韧性、塑性和延展性显著下降。硅可以使材料的屈服极限和弹性极限得到显著提高, 对疲劳强度也有较好的改善。

钒　钒是强碳化物形成元素, 在合金中首先与碳结合形成碳化钒 (VC), 对提高材料的耐磨性极为有利。为了析出足够多的VC, 减少残余奥氏体量, 提高合金铸态时的硬度, 应加入足够多的钒。当V含量低于9%时, 初生碳化钒形态主要为团球状及块状。随合金中钒加入量的增加, VC也慢慢地聚集变得粗大。铸态时, 碳与钒结合既生成初生碳化物, 又生成二次碳化物。VC的显微硬度可达2 300~2 800 HV。钒在铸铁中主要以三种状态分布: ①固溶于α-Fe中。②析出相。③块状化合物。[7]

铬　铬与碳的结合能力比铁和锰强, 它与碳可以形成不同类型的碳化物。它可以将渗碳体中部分铁原子置换出来而形成 (Fe, Cr)$_3$C的含铬合金渗碳体。这种M$_3$C型碳化物的硬度在1000~1230 HV。当材料中铬含量较高时, 可以形成 (Cr, Fe)$_{23}$C$_6$和 (Cr, Fe)$_7$C$_3$的复杂碳化物, 其硬度分别为1140 HV和1300~1800 HV。铬含量的多少可以影响奥氏体区域的大小。铬还可以少量地溶于奥氏体中, 提高基体的淬透性, 并且淬透性随Cr/C的增加而提高。基体中铬的质量分数Cr%可以用公式 (2-2) 估算出来。

$$Cr\% = \left(\frac{1.95Cr}{C} - 2.47 \right)\% \tag{2-2}$$

锰　锰能扩大γ相区, 是稳定奥氏体元素。当合金中的钛、锆、钒、铌形成难溶于奥氏体的碳化物时, 会降低材料的淬透性。加入锰可以使这些难溶碳化物元素的不利影响发生改变, 提高材料的淬透性。[8]同时, 形成的Fe-Mn-C原子团, 诱变奥氏体转变为马氏体, 使合金具有加工硬化能力。[9]

在合金元素成分设计上, 本实验主要考虑以下3个因素: ①得到数目、形态和分布较为合理的碳化物 (包括初生碳化物和共晶碳化物)。作为参考所涉及的各类合金元素碳化物的硬度范围在Table5-2-1中列出。②为了使基体具有一定的强度来支撑和镶嵌碳化物。③为了使晶粒细化。同时, 本课题设计的耐磨材料主要用于磨粒磨损工况, 耐磨件通常都要承受一定的冲击。要求耐磨件在具有高耐磨性的同时, 必须具有足够的韧性。因此, 以冲击磨损工况为应用目标的新型多元合金, 在成分设计上有两个基本要求: 一是要有大量的高硬度的硬质点, 使合金具有较高的耐磨性; 二是碳化物的形态要好, 分布尽量均匀, 冲击韧性较高。综合以上几个因素, 并参考之前研发材料的成果及设计思想, 本课题设计的新型Fe基F-V-Cr-Mn合金材料, 铸态下用Fe-V-Cr-Mn表示, 成分见Table5-2-2。

Table 5-2-1　The Vickers hardness of some carbides

碳化物	Fe$_3$C	(Cr, Fe)$_3$C	(Cr, Fe)$_{23}$C$_6$	(Cr, Fe)$_7$C$_3$	Mo$_2$C	VC
硬度 (HV)	900~1 000	1 000~1 230	1 140	1 300~1 800	1 500	2 300~2 600

Table 5-2-2　Chemical composition of newly developed Fe-based Fe-V-Cr-Mn alloy　　(wt%)

C	Si	Mn	Cr	V
2.3	1.0	5.0	8.0	7.0

5.2.2.2　实验原材料

使用的原材料主要为高碳铬铁、钒铁、硅铁、高碳锰铁、低碳锰铁、生铁、废钢、纯镍板等。其主要成分为Table5-2-3, Table5-2-4, Table5-2-5, Table5-2-6, Table5-2-7, Table5-2-8中所示。

Table 5-2-3　Chemical composition of high carbon ferro-chrome　　　　　（wt%）

C	Si	Cr	S	P
8	1.89	59.01	0.026	0.144

Table 5-2-4　Chemical composition of ferro-vanadium　　　　　（wt%）

C	Si	Mn	V	S	P	Al
0.53	0.61	0.43	49.38	0.032	0.058	0.01

Table5-2-5　Chemical composition of ferro-silicon　　　　　（wt%）

C	Si	S	P	Al
0.15	72.1	0.016	0.034	1.49

Table 5-2-6　Chemical composition of high carbon ferromanganese　　　　　（wt%）

C	Si	Mn	S	P
6.5	0.8	65.1	0.02	0.17

Table 5-2-7　Chemical composition of pig iron　　　　　（wt%）

C	Si	Mn	S	P
4.37	0.68	0.12	0.026	0.05

Table 5-2-8　Chemical composition of scrap steel　　　　　（wt%）

C	Si	Mn	S	P
0.2	0.4	0.45	0.1	0.04

5.2.2.3　实验设备

本次实验所用到的所有设备在Table5-2-9中列出。

Table 5-2-9　The equipments used in the present study

设备名称	型号与参数
中频感应炉	60 kW, 650 V
光学显微镜	日本OLYMPUS-GX51
能谱分析仪	HORIBA, 7021-H型
X射线衍射仪	日本理学D/MAX-2500/PC
数字维氏硬度计	HVS-30Z\LCD
磨损试验机	日本SUGA NUS-IS03
激光共聚焦显微镜	德国卡尔·蔡司LSM-700

5.2.2.4　制备过程

本实验中利用中频感应电炉，对原材料如废钢、生铁、钒铁、铬铁等进行熔炼、制备合金。在之前研发材料的基础上确定熔炼方法，制备出三种成分的新型含多元碳化物合金复合材料，设计其基体主要为奥氏体，析出碳化物为VC，M_7C_3等，并将制备两种国内外常规耐磨材料（高铬铸铁、高锰钢）作为对比材料。

实验详细过程为：本实验在30 kg的中频感应炉（电炉额定容量（钢）30 kg，额定功率：60 kW，电源额定输出电压（单相）：650 V，输出频率：2500 Hz，熔炼时间：40 min，倾炉速度：自由手动，冷却水流量3 m³/h。）中熔化炉料，炉料分别按废钢—生铁—高碳铬铁—高碳锰铁—钒铁—硅铁的顺序加入。熔炼温度为1 650~

1 700℃，出炉前使溶液在此温度下静置保温10~15 min，最后出炉浇注，浇注温度为1 550~1 600℃。出炉前用铝丝终脱氧。使用黏土砂造型，在砂型内侧涂抹石墨粉，减少铸件粘砂。模样为板状试块和Y型试块，如Fig.5-2-1所示。

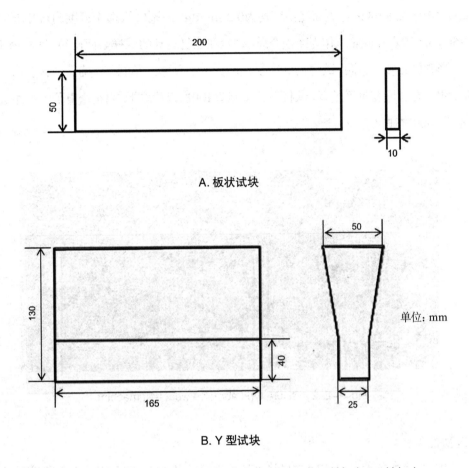

A. 板状试块

B. Y 型试块

Fig. 5-2-1　Shape and dimensions of plate-shaped and Y-shaped blocks

用浇铸出来的Y型试块下方40 mm×25 mm×165 mm的部分通过线切割，制成标准的无缺口冲击试验件，其尺寸为10 mm×10 mm×55 mm，板状试块切割为50 mm×50 mm×4 mm的磨损实验片。

5.2.3　实验方法

5.2.3.1　金相组织观察

通过金相组织观察，可以分析合金中基体的形貌特征及碳化物的形态、分布、大小等，这是分析组织与性能之间联系的重要环节。

将试样加工成10 mm×10 mm×10 mm的尺寸，经砂轮机上粗磨打倒角→不同倍数下砂纸粗磨→粗抛布粗磨→精抛布抛光→腐蚀后利用蔡司Axiovert 40型金相显微镜（德国），观察合金50×，100×，200×，500×，1000×倍下的金相组织。

5.2.3.2　磨粒磨损实验

本文所使用的耐磨性评估方法为磨粒磨损实验。它可以直观地反应材料本身抗磨粒磨损性能的强弱,也可以直接比较几种材料的抗磨粒磨损性能大小。

本实验使用的SUGA型磨粒磨损试验机,其外部构造如Fig.5-2-2所示。磨损试样尺寸为50 mm×50 mm×4 mm,测试载荷为19.6N,测试试样表面光洁度均为0.2 μm Ra。测试砂纸选用粒度为180目的碳化硅砂纸,尺寸为12 mm×158 mm,固定在砂轮上作为磨损介质。砂轮在往复运动的试样表面旋转,试样每往复一个来回,砂轮旋转0.9°。当砂轮旋转一整圈,试样正好往复运动400次,更换一次砂纸,确保试样表面一直接触到全新的砂纸的磨损,测试在相同变量下进行。试样每往复运动100次以后取下,用精密电子天平测量一次磨损量。每个试样都进行2 000次往复磨损。

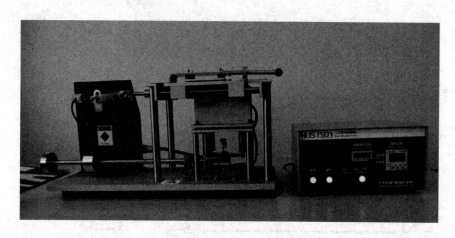

Fig. 5-2-2   SUGA type abrasive wear test machine

5.2.3.3   冲击韧性实验

在实验中,用摆锤冲击弯曲试验来测定材料抵抗冲击载荷的能力,即测定冲击载荷试样被折断而消耗的冲击功A_k,单位为焦耳(J)。而用试样横截面积F(有缺口的为缺口处的横截面积)去除A_k,可得到材料的冲击韧度(冲击值)指标,即公式(2-3),其单位为 kJ/m² 或 J/cm²。

$$a_k = A_k / F \tag{2-3}$$

将每组试样线切割成三个10 mm×10 mm×55 mm无缺口的标准试样(高锰钢试样切U型口),冲击韧性取其平均值。

5.2.4   实验表征方法

5.2.4.1   硬度测试

硬度值作为衡量材料耐磨性的重要指标之一,可以一定程度地反应材料耐磨性能的强弱。

材料硬度试验按国标GB/T230-91《金属洛氏硬度试验方法》进行测量。表面硬度测量在HR-150A型洛式硬度计上进行,载荷选取150 kg。每个试样取5个点,最后取平均值,即为试样的硬度值(HRC)。

5.2.4.2   显微硬度

为确定试样组织中基体及碳化物的硬度,在HVS-30Z型数字显微硬度计上进行显微硬度的测定。将照完金相的试样在0.1 kg的载荷下加载20 s后进行测量。

5.2.4.3　X射线衍射分析

X射线衍射仪可以对材料进行物相分析,确定合金的基体及碳化物相。

将金相试样精磨抛光后不腐蚀,利用X光衍射仪(D/MAX-2500,日本理学)进行材料分析,利用分析结果初步判断材料中碳化物及其基体的组成。

5.3　铸态新型Fe基Fe-V-Cr-Mn合金的耐磨性能

5.3.1　引言

高锰钢自发明以来,因为它具有良好的加工硬化性能,以及在高冲击工况下表现出来的高强度、高塑性、韧性好、特别耐磨等特点,被广泛用于冶金、建材、电力、矿山、铁路、煤炭、军工、农机等各个领域中。[9-11]

高铬铸铁是第三代抗磨材料,由于其自身的组织特点,具有比普通合金钢和白口铸铁的耐磨性、耐热性、强度及韧性高得多。铬系铸铁磨球近几年已经取代了部分低合金钢、中锰球铁等材质的磨球,被认为是目前较好的耐磨材料之一。[12-14]

新型Fe基Fe-V-Cr-Mn合金中含有V,Cr等合金元素。钒是强碳化物形成元素,首先与碳结合形成MC型碳化物从液相中析出,在冷却过程中又在基体中析出二次碳化物,基体上形成高硬度的碳化物相。同时合金中含有Mn元素,可以增强基体的形变硬化能力,使之与普通耐磨材料相比,具有更好的耐磨性及韧性。

本章将对铸态新型Fe基Fe-V-Cr-Mn合金与高锰钢、高铬铸铁这两种成熟的耐磨材料进行耐磨性能及冲击韧性的比较,然后分析结果,得出结论。

5.3.2　实验材料及金相组织

本章所用到的材料为Fe-V-Cr-Mn、高锰钢(以下用Mn13Cr2表示)与高铬铸铁(以下用Cr27Mo1表示)。实验材料的化学成分对比见Table5-3-1,

Table5-3-1　Chemical composition of test samples　　　　　　　　(wt%)

Samples	C	Si	V	Mn	Cr	Mo	P	S
Fe-V-Cr-Mn	2.3	1.0	7.0	5.0	8.0	—	—	—
Cr27Mo1	2.7	0.5	—	0.55	27	1	—	—
Mn13Cr2	1.25	0.7	—	13	2	—	≤0.06	≤0.04

将三种材料进行制样、研磨、抛光,分别观察三种材料的金相组织,如Fig.5-3-1所示,其中Fe-V-Cr-Mn

用浓盐酸和浓硝酸体积比按3:1制成王水进行腐蚀。Fig.5-3-1a为Fe-V-Cr-Mn合金的显微组织，Fig.5-3-1b为Mn13Cr2的显微组织，Fig.5-3-1c为Cr27Mo1的显微组织。

Fig. 5-3-1　Metallographic structures of different test materials
（a: Fe-V-Cr-Mn, b: Mn13Cr2, c: Cr27Mo1）

Fe-V-Cr-Mn经过王水腐蚀后，它使合金基体呈现为黑色，碳化物呈现白色和灰色，如Fig.5-3-1a所示。通过EDS（Fig.5-3-2）分析可知，V主要集中聚集在一些块状和颗粒状区域，通过显微硬度计测量，硬度为2427HV。Cr主要集中在一些断裂的网状及短条状聚集处，测量其显微硬度为1 143 HV。Mn均匀地分布在基体上，同时发现，Fe主要分布于基体及富铬的区域，在富V的区域内，Fe的含量极少。通过Fe-V-Cr-Mn的X射线衍射分析（Fig.5-3-3c）可知，V的存在形式主要是以MC型的VC为主，其形状为块状、颗粒状。Cr的存在形式主要是以Cr23-XFeXC6为主，它是由许多短条状聚集在一起，类似于一些断续的网状结构。Fe-V-Cr-Mn基体组织主要由奥氏体+铁素体组成，显微硬度为589 HV左右。本实验用到的对比材料Mn13Cr2中含有2%的Cr，它既溶于铁素体中，提高钢的淬透性，又可以与Fe，C形成网状碳化物，在晶界上析出，如Fig.5-3-1b所示。经分析，水韧处理后的高锰钢中基体组织主要为奥氏体组织（Fig.5-3-3a），显微硬度为305 HV左右。本实验中用到的另外一种对比材料Cr27Mo1经过淬火+回火的热处理后，基体组织主要由马氏体和残余奥氏体组成（Fig.5-3-3b），显微硬度为895 HV左右。由于合金中加入了1%的Mo，所以组织中不但有连续的一次碳化物出现（显微硬度为1 450 HV），在基体上也分布着更为细小的二次碳化物颗粒（Fig.5-3-1c）。

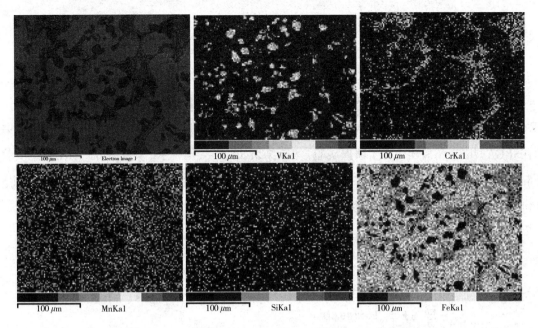

Fig. 5-3-2　EDS elements distribution of Fe-V-Cr-Mn

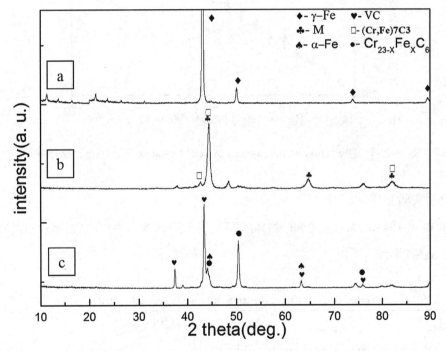

Fig. 5-3-3　X-ray diffraction spectrum of three different materials
（a: Mn13Cr2, b: Cr27Mo1, c: Fe-V-Cr-Mn）

　　另外，本实验应用Image-Pro Plus软件（IPP）对材料中碳化物相面积分数进行了计算。先通过Photoshop软件调节金相图片的对比度、色阶等，使图像处理后的金相组织照片中基体与碳化物明显区分开来，然后通过IPP软件对它们分别进行染色标定，最终计算确定Fe-V-Cr-Mn合金中碳化物所占面积比例为20%~25%。

5.3.3　耐磨性能及冲击韧性对比

5.3.2.1　SUGA 型磨损试验机重复性

本实验用到的磨损试验机为SUGA型磨粒磨损试验机（详见第2章）。

在实验前，为了验证SUGA型磨粒磨损试验机的实验可重复性，用同种材料做了三个完全相同的磨损试验片，分别在SUGA型磨损试验机上做了磨损实验，实验结果如Fig.5-3-4所示。从图中可以看出三组实验磨损率基本相同，曲线基本重合，可以得出SUGA型磨损试验机的实验重复性很高。在进行所有磨损试验前，通过处理，保证每个磨损试验片的表面粗糙度相同。

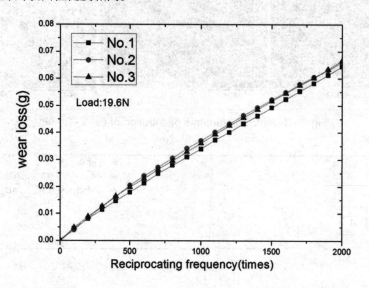

Fig. 5-3-4　The repeated experiment of SUGA type wear testing machine

5.3.2.2　磨粒磨损实验结果

本实验分别对Fe-V-Cr-Mn，Mn13Cr2和Cr27Mo1进行了表面硬度测试及SUGA型磨粒磨损实验，实验结果如Table.5-3-2，Fig. 5-3-5所示。

Table5-3-2　The hardness of different materials（HRC）

材料	硬度值（HRC）					平均值
	1	2	3	4	5	
Fe-V-Cr-Mn	47.5	48	47.5	47.9	48	47.8
Mn13Cr2	20.5	23.7	19.4	20.2	22.5	21.3
Cr27Mo1	65.1	64. 4	67.2	66.5	65.0	65.6

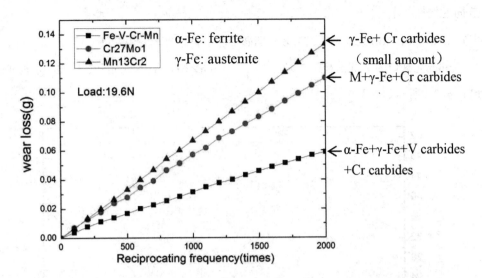

Fig.5-3-5　Abrasion results of different materials

从Table5-3-2可以看出，Fe-V-Cr-Mn的硬度在铸态下达到了47.8 HRC，比Cr27Mo1的硬度低27%，比未磨损前的Mn13Cr2硬度高55%。

三种材料经过SUGA型磨料磨损试验机的磨损后，Fe-V-Cr-Mn磨损表面显微硬度经测试为744 HV，比磨损前增加了26%；Mn13Cr2磨损表面的显微硬度经测试为628 HV，比磨损前增加了106%；Cr27Mo1磨损表面的显微硬度经测试为900 HV，磨损前后的表面显微硬度基本没有变化，如Fig.5-3-6所示。

69

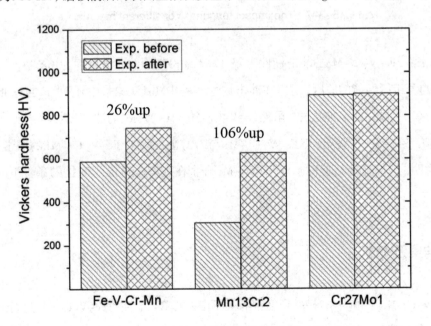

Fig.5-3-6　Change of Vickers hardness for different materials after test

磨损往复次数与磨损量之间的关系如Fig.5-3-5所示。从图中可以看出，虽然Fe-V-Cr-Mn的硬度不及Cr27Mo1，但是在相同的磨损往复次数下，Fe-V-Cr-Mn的磨损量却最少，其次为Cr27Mo1，Mn13Cr2的磨损量在本次实验中最多。通过最终计算，Fe-V-Cr-Mn合金的磨损量约为Cr27Mo1的1/2，Mn13Cr2的2/5；设计的新型Fe基Fe-V-Cr-Mn的耐磨性约为高铬铸铁的两倍，高锰钢的2.5倍。

5.3.2.3 冲击韧性对比

本实验对Fe–V–Cr–Mn, Mn13Cr2和Cr27Mo1分别做了冲击韧性实验, 其中, Fe–V–Cr–Mn和Cr27Mo1均采用10 mm×10 mm×55 mm无缺口的标准试样, Mn13Cr2采用10 mm×10 mm×55 mm开U型口的标准试样, 三者的冲击韧性对比如Fig.5–3–7所示。

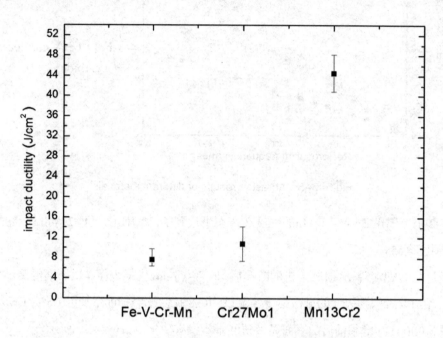

Fig. 5–3–7　The impact toughness of different materials

从实验结果来看, Fe–V–Cr–Mn的冲击韧性为7.8 J/cm², Cr27Mo1的冲击韧性为10.7 J/cm²。

由于高锰钢自身组织性能的原因, 所以在冲击韧性实验中Mn13Cr2采用开U型口的标准试样, Fe–V–Cr–Mn和Cr27Mo1均采用无缺口的标准试样。虽然Fe–V–Cr–Mn与Cr27Mo1的冲击韧性值不能直接与Mn13Cr2进行比较, 不过从Fig.5–3–8可以看出, 就Fe–V–Cr–Mn与Cr27Mo1而言, Fe–V–Cr–Mn的冲击韧性值略微低于Cr27Mo1。结果说明, 所设计的新型Fe基Fe–V–Cr–Mn合金在拥有较高的耐磨性的条件下, 仍然保持着良好的冲击韧性。

5.3.4　实验结果及分析

磨损后, Fe–V–Cr–Mn与Mn13Cr2显微硬度都有大幅度的提高。单独对Fe–V–Cr–Mn进行磨损前后的X射线衍射分析, 如Fig.5–3–8所示。Mn13Cr2中存在Fe–Mn–C原子团, 在磨损时由于发生加工硬化现象, 形变诱发奥氏体转变为马氏体。通过分析发现, 在Fe–V–Cr–Mn中, 也同样发生了类似于高锰钢加工硬化的现象, 结果导致合金中部分的奥氏体转变为马氏体, 合金显微硬度提高, 耐磨性增加。

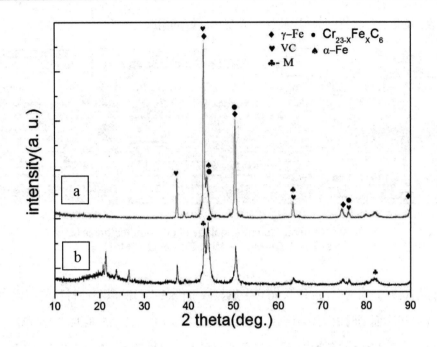

Fig. 5-3-8　X-ray diffraction analysis on Fe-V-Cr-Mn before and after wear
(a: before test, b: after test)

　　为了更进一步地分析这三种实验材料出现耐磨性与冲击韧性差异的原因，本实验使用光学显微镜观察了三种材料磨损后的试样表面，Fig.5-3-9为三种材料放大50倍后的磨损表面。从图中可以看出，磨损面总体上都比较平整，但在Mn13Cr2磨损表面出现了非常明显的裂纹和孔洞。

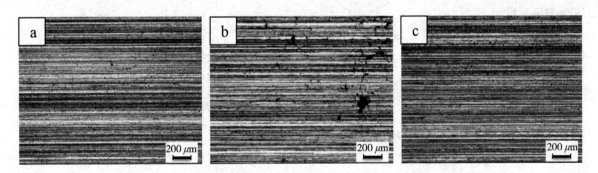

Fig. 5-3-9　Worn surface morphology of different materials (50×)
(a: Fe-V-Cr-Mn, b: Mn13Cr2, c: Cr27Mo1)

　　利用光学显微镜在500倍下重新观察三种材料磨损后的表面形貌，如Fig.5-3-10所示，经过与砂纸一定次数的往复磨损后，三个试样表面均形成犁沟。从Fig.5-3-10b中还可以更加清晰地看到Mn13Cr2表面由于材料剥落导致的孔洞及周围的裂纹。同时从Fig.5-3-10c中也发现，在Cr27Mo1的磨损表面也出现了细微的裂纹，还有比较深的犁沟。对比三种材料在500倍下的磨损照片，从直观上看不管是划痕数量还是深度，Fe-V-Cr-Mn都要小于Cr27Mo1与Mn13Cr2。

　　材料在磨粒磨损进程中产生犁沟的过程并不会直接引起材料的去除，但在多次变形后会产生脱落而形成二次切屑。Fig.5-3-10a中箭头所指处，划痕在经过块状VC时被阻断或减弱。这种阻断既缩短了犁沟的长度，又减少了变形的次数, 降低了对材料的磨损, 使材料表现出良好的耐磨性。

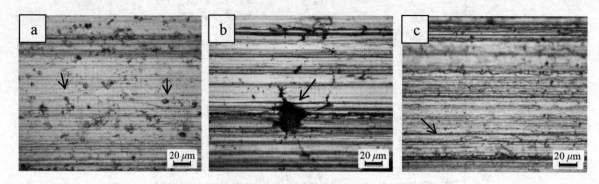

Fig.5-3-10　Worn surface morphology of different materials（50×）
（a: Fe-V-Cr-Mn, b: Mn13Cr2, c: Cr27Mo1）

　　为了进一步详细观察磨损情况, 利用激光共聚焦显微镜LSM-700模拟三种材料的磨损表面三维形貌, 如Fig.5-3-11所示, 在磨损表面同时放大100倍后, 可以发现Fe-V-Cr-Mn（Fig.5-3-11a）与Cr27Mo1（Fig.5-3-11c）的磨损表面起伏都是在0~150 μm内变化, 但从颜色代表的起伏高度来看, Fe-V-Cr-Mn中表示起伏高度在140 μm以上的深色区域要明显少于Cr27Mo1的, 意味着Fe-V-Cr-Mn的磨损程度要比Cr27Mo1低, 具有更好的耐磨性。Mn13Cr2（Fig.5-3-11b）经过磨粒磨损后, 虽然磨损表面起伏高度并没有其他两种材料的大, 都是在0~100 μm内变化, 但由于表面有材料发生了剥落, 使得高锰钢宏观表现为材料损耗最多, 耐磨性最差。

　　另外, 本实验使用线切割将磨损试件的磨损表面平行于磨损方向及垂直于磨损方向切割, 然后从方向①观察平行于磨损方向的磨损表面变化情况, 从方向②观察垂直于磨损方向的磨损表面变化情况, 试样切割示意图如Fig.5-3-12所示。

72

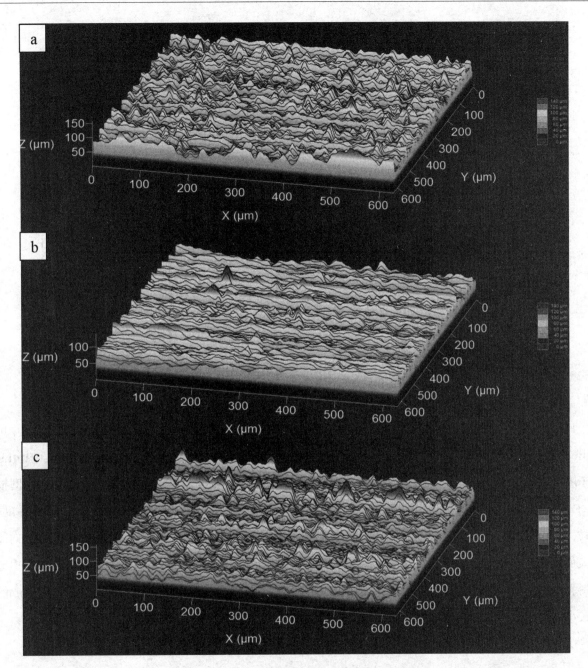

Fig. 5-3-11　3D pictures comparison of worn surface（100×）
（a: Fe-V-Cr-Mn，b: Mn13Cr2，c: Cr27Mo1）

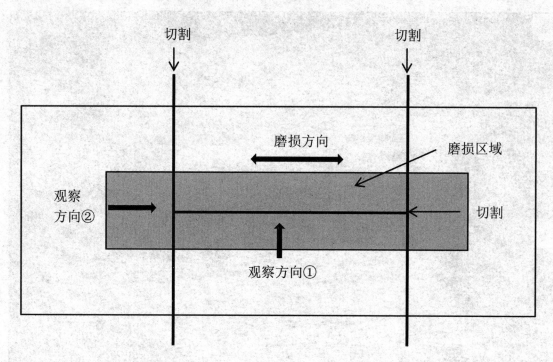

Fig. 5-3-12   The Schematic diagram of sample cutting

　　Fig.5-3-13是三种材料平行于磨损表面切割后从方向①观察表面变化的金相照片（500×）。从图中可以看出，三种材料的磨损表面均发生了由于基体变形产生的变形磨损及磨粒磨损产生的切削磨损，其中Mn13Cr2由于磨损前的硬度很小，主要是以变形磨损为主；Cr27Mo1的硬度较大，在磨损过程中主要是以切削磨损为主；Fe-V-Cr-Mn的变形磨损与切削磨损都介于两者之间。在磨损过程中，Fe-V-Cr-Mn中的球状碳化物即VC颗粒，表现出了超高的硬度及与基体紧密的结合力。Fig.5-3-14是三种材料垂直于磨损表面切割后从方向②观察表面变化的金相照片（500×）。从Fig.5-3-14b中可以明显地看出Mn13Cr2在磨损过程中由于变形磨损导致材料剥落后留下的孔洞。Fe-V-Cr-Mn与Cr27Mo1在此切面下看到的磨损表面都有所起伏，但Fe-V-Cr-Mn的起伏较小。

a. Fe-V-Cr-Mn

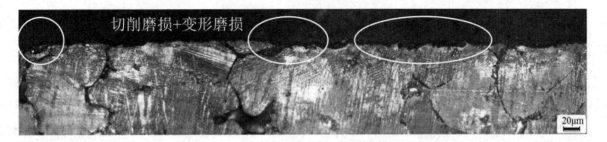

b. Mn13Cr2

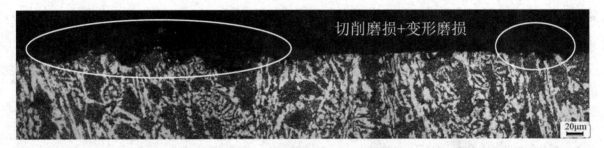

c. Cr27Mo1

Fig. 5-3-13　Cross sectional metallographic photos parallel to wear direction（observation direction ①）

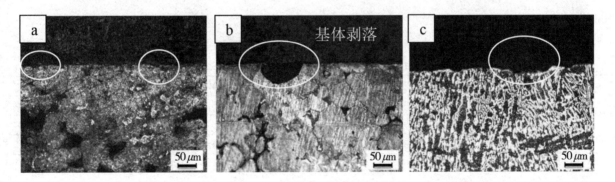

Fig. 5-3-14　Cross sectional metallographic photos perpendicular to
wear direction（observation direction ②）
（a: Fe-V-Cr-Mn, b: Mn13Cr2, c: Cr27Mo1）

5.4 改性Fe基Fe-V-Cr-Mn合金的耐磨性能

5.4.1 引言

稀土作为极为重要的战略资源,素有"工业维生素"之美誉。中国是世界公认的稀土大国,拥有最多的稀土资源储量及最大的稀土产量。

稀土加入合金材料中,变质处理可以起到细化晶粒、强化晶界的作用,同时也具有改善碳化物的形态、脱氧、脱硫、净化铁液的作用。

另外,镍(Ni)是一种稀缺资源,也是一种重要的战略物资。一定含量的Ni可以在不显著降低钢的韧性的情况下提高其强度。Ni在提高钢强度的同时,对钢的塑性、韧性以及其他工艺性能的损害较其他合金元素要小。合金中Ni主要溶于基体,不溶于碳化物。Ni可以扩大奥氏体相区,是稳定奥氏体的主要元素。同时,合金中加入一定量的Ni,能有效地提高淬透性,在常温下基体中会有少量的残余奥氏体剩余,其韧性会提高。此外,合金中加入Ni,可以提高其耐酸碱腐蚀性,Ni是不锈耐酸钢中的重要元素之一。

本节主要介绍Fe-V-Cr-Mn中分别添加0.2%1#稀土(RE: 28.4%, Si: 39.6%, 余量Fe)(以下用Fe-V-Cr-Mn-RE表示)和3%Ni(以下用Fe-V-Cr-Mn-Ni表示)后铸态下对新型Fe基Fe-V-Cr-Mn合金的组织与性能的影响。

5.4.2 实验材料及金相组织

本章所用到的材料为Fe-V-Cr-Mn, Fe-V-Cr-Mn-RE, Fe-V-Cr-Mn-Ni三种合金。三种合金的化学成分见Table5-4-1。

Table 5-4-1 Chemical composition of test samples (wt%)

Samples	C	Si	V	Mn	Cr	RE	Ni
Fe-V-Cr-Mn	2.3	1.0	7.0	5.0	8.0	—	—
Fe-V-Cr-Mn-RE	2.3	1.0	7.0	5.0	8.0	0.2	—
Fe-V-Cr-Mn-Ni	2.3	1.0	7.0	5.0	8.0	—	3

将三种不同成分的合金进行制样、研磨、抛光,用王水腐蚀后分别观察它们的金相组织,如Fig.5-4-1所示。在铸态下Fe-V-Cr-Mn加入0.2%的稀土后,合金中V的碳化物尺寸明显变小,数量增加,但对Cr的碳化物并没有多少影响,如Fig.5-4-1b所示。同样使用Image-Pro Plus软件测得Fe-V-Cr-Mn-RE中碳化物所占面积比例为15%~18%,少于Fe-V-Cr-Mn中的碳化物面积比例。通过EDS分析可知,稀土Ce主要存在于VC相中(Fig.5-4-2)。溶解的稀土起到微合金化的作用,使合金中VC球化、细化,并且呈一定程度的集聚。Fe-V-Cr-Mn-Ni中由于加入了Ni元素,使得合金的耐酸腐蚀变强,经王水腐蚀后,金相组织在显微镜下的颜色较其他两种合金仍比

较明亮（Fig.5-4-1c）。通过EDS分析可知，Ni均匀地分布在合金基体上，对V，Cr的碳化物并没有造成多大的影响（Fig.5-4-3）。

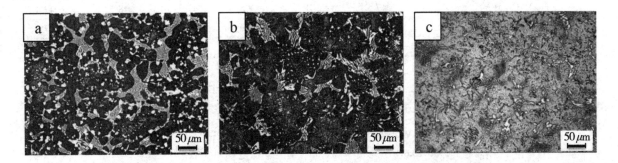

Fig. 5-4-1　Metallographic structures of different test materials
（a：Fe–V–Cr–Mn，b：Fe–V–Cr–Mn–RE，c：Fe–V–Cr–Mn–Ni）

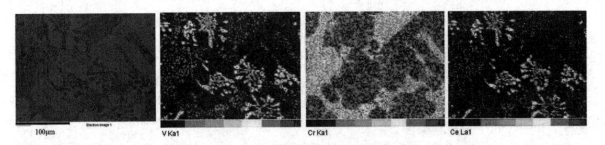

Fig. 5-4-2　The distribution of V, Cr and Ce in Fe–V–Cr–Mn–RE

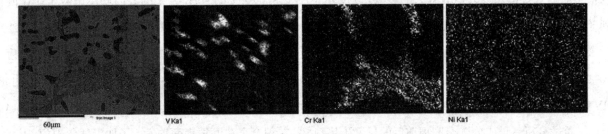

Fig. 5-4-3　The distribution of V, Cr and Ni in Fe–V–Cr–Mn–Ni

5.4.3　耐磨性能及冲击韧性对比

5.4.3.1　磨粒磨损实验结果

　　本章对Fe–V–Cr–Mn，Fe–V–Cr–Mn–RE和Fe–V–Cr–Mn–Ni三种合金分别进行了硬度测试及SUGA型摩擦磨损实验，实验结果见Table5-4-2，Fig.5-4-4。

Table5-4-2 The hardness of different materials（HRC）

材料	硬度值（HRC）					平均值
	1	2	3	4	5	
Fe-V-Cr-Mn	47.5	48	47.5	47.9	48	47.8
Fe-V-Cr-Mn-RE	47.7	47.9	47.9	45.8	47.5	47.4
Fe-V-Cr-Mn-Ni	41.2	40.7	41	40.2	39.6	40.5

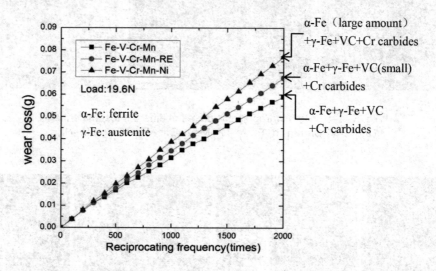

Fig. 5-4-4 Abrasive wear results of three different materials

从Table5-4-2可以看出，Fe-V-Cr-Mn的硬度平均值为47.8 HRC，加入稀土后的表面硬度为47.4 HRC，加入稀土后对合金的硬度并没有太大的影响，硬度基本不变。加入3%的Ni后，合金的表面硬度由47.8 HRC降低到40.5 HRC，降低了15%。这是由于加入Ni后，合金的基体中铁素体（α-Fe）的含量增加。合金中各种基体组织显微硬度见Tabble5-4-3。由Table5-4-3可知，奥氏体（γ-Fe）的显微硬度为300~600 HV，铁素体的显微硬度为70~200 HV，所以铁素体含量的升高会导致基体硬度的下降。磨损后，Fe-V-Cr-Mn-RE基体的显微硬度由585 HV变为719 HV，提高了23%。Fe-V-Cr-Mn-Ni基体的显微硬度由385 HV变为494 HV，提高了28%，如Fig.5-4-5所示。通过分析三种成分合金磨损前后的显微硬度可知，三种合金在磨损表面均发生了加工硬化现象，其中，Fe-V-Cr-Mn-Ni磨损后基体的硬度增幅最大。三种成分的合金相同磨损后基体硬度最大的是Fe-V-Cr-Mn。

三种不同成分的合金在19.6 N的载荷下，经过SUGA磨料磨损试验机的磨损后，磨损往复次数与磨损量之间的关系如Fig.5-4-4所示。结果显示，三种合金的磨损曲线都比较平整，从磨损量来看，Fe-V-Cr-Mn-RE比Fe-V-Cr-Mn合金多15%，Fe-V-Cr-Mn-Ni比Fe-V-Cr-Mn合金多31%。从结果来看，分别加入稀土和加入Ni后合金的磨损量增加，耐磨性反而下降，这与这些元素加入后使合金组织形态发生变化有关。

Table5-4-3 The Vickers hardness of different matrix

基体组织	铁素体	珠光体	奥氏体	马氏体
硬度（HV）	70~200	300~460	300~600	500~1000

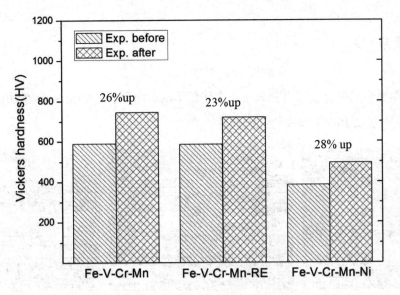

Fig. 5-4-5　Change of Vickers hardness for different materials after test

5.4.3.2　冲击韧性对比

本章对Fe-V-Cr-Mn-RE和Fe-V-Cr-Mn-Ni两种不同成分的合金分别做了冲击韧性实验,并与Fe-V-Cr-Mn进行对比。冲击试样均制成10 mm×10 mm×55 mm的无缺口标准试样,三者的冲击韧性对比如Fig. 5-4-6所示。

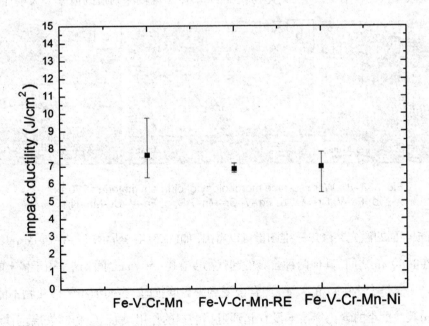

Fig. 5-4-6　The impact toughness of different materials

从实验结果来看,分别加入稀土和Ni后,并没有使合金的冲击韧性值得到提高,相反,还有略微下降。

5.4.4 实验结果及分析

应用光学显微镜观察三种合金磨损表面在放大50倍下的表面形貌，如Fig.5-4-7所示。从图中可以看出，三种合金的磨损表面在放大50倍下表现的都比较平整。

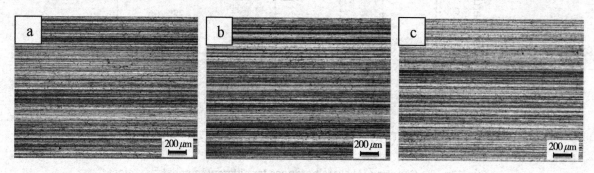

Fig. 5-4-7 Worn surface morphology of different materials（50×）
（a: Fe-V-Cr-Mn, b: Fe-V-Cr-Mn-RE, c: Fe-V-Cr-Mn-Ni）

用光学显微镜在放大500倍下重新观察三者的磨损表面，如Fig.5-4-8所示，观察发现Fe-V-Cr-Mn的磨损表面比其他两种合金的磨损表面起伏较剧烈，表现的并没有其他两种合金平整。在三种合金中，凡是碳化物密集的区域，划痕均较少且比较细微；而在碳化物较少的地方，出现较深划痕的概率大大增加。

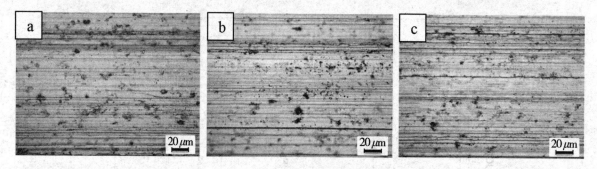

Fig. 5-4-8 Worn surface morphology of different materials（500×）
（a: Fe-V-Cr-Mn, b: Fe-V-Cr-Mn-RE, c: Fe-V-Cr-Mn-Ni）

从三种合金磨损表面的激光共聚焦三维图像可以看出，如Fig.5-4-9所示，Fe-V-Cr-Mn-RE的表面相对起伏高度是最小的，在0~80 μm之间，其他两种合金的起伏高度在0~150 μm之间。这是由于稀土的变质作用使得碳化物的尺寸变小，尤其是合金中VC的尺寸。当合金受到外界磨损时，表面细小的碳化物不能很好地阻挡磨粒的磨损，甚至部分尺寸更小的碳化物根本没有起到阻挡磨损的作用，就已经随同表层基体一同被脱落，所以从同一倍数下看Fe-V-Cr-Mn-RE磨损后的表面比Fe-V-Cr-Mn磨损表面表现得较为平整。

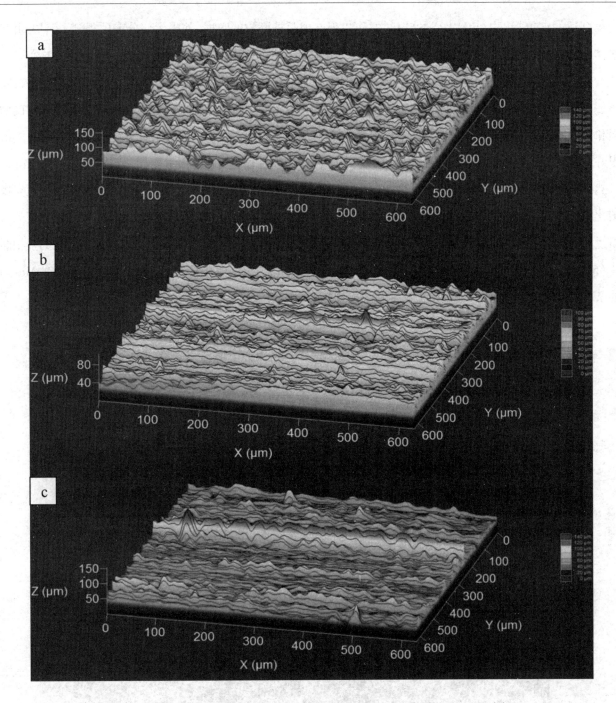

Fig. 5-4-9 Comparison of 3D images of worn surface of different materials
（a: Fe-V-Cr-Mn, b: Fe-V-Cr-Mn-RE, c: Fe-V-Cr-Mn-Ni）

81

当合金表面某一区域内由于碳化物与基体的共同作用使得这一区域抵抗磨损的能力较它两侧的区域强时，两侧区域表面磨损较多凹陷下去，这一区域就凸显出来，形成了连续的高峰。Fe-V-Cr-Mn-Ni的磨损表面起伏高度与Fe-V-Cr-Mn相当；但在Fe-V-Cr-Mn中，起伏高度基本位于$80~\mu m$左右，且表面起伏较剧烈。在Fe-V-Cr-Mn-Ni中，有大片的区域起伏高度位于$60~\mu m$左右，表面较平整，但有连续的高峰出现，这说明在Fe-V-Cr-Mn-Ni中有较多区域的基体在磨损中剥落，这是因为合金加入Ni后，表面硬度下降，基体变软，容易磨损剥落。

所以，Fe-V-Cr-Mn虽然从磨损表面三维图像上看表面起伏较大，但对比其他两种成分的合金，由于含有大小合适的V的碳化物（VC）及Cr的碳化物，可以很好地阻挡磨粒的磨损，使得大部分表层得到了保护，并且基体上具有一定的硬度，也可以反过来支撑这些碳化物的存在，所以宏观表现为磨损量较少，耐磨性较好。

对三种材料磨损表面同样做了平行于磨损方向及垂直于磨损方向的切割。Fig.5-4-10是三种材料平行于磨损表面切割后从方向①观察表面变化的金相照片（500×），可以看出三种材料的磨损表面都发生了切削磨损和变形磨损。其中Fe-V-Cr-Mn-RE中由于稀土的变质作用使得VC硬质相的尺寸比变质前变少了很多，不能很好地起到保护基体的作用，随同基体一同被剥落，使得磨损过程中切削磨损量变大。在Fe-V-Cr-Mn-Ni中，虽然碳化物的数量和尺寸没有明显的变化，但由于加入Ni后基体的硬度下降较多，基体不能起到很好地支撑碳化物的作用，所以在磨损过程中，基体容易首先被破坏，对碳化物的保护作用减弱，使得磨损过程中切削磨损量与变形磨损量同时增大。Fig.5-4-11是三种材料垂直于磨损表面切割后从方向②观察表面变化的金相照片（500×）。从图中可以看出，虽然三者的磨损表面总体上都较平整，但在Fe-V-Cr-Mn-RE和Fe-V-Cr-Mn-Ni的某些磨损表面区域内，切削磨损量及变形磨损量都要比Fe-V-Cr-Mn大很多。

82

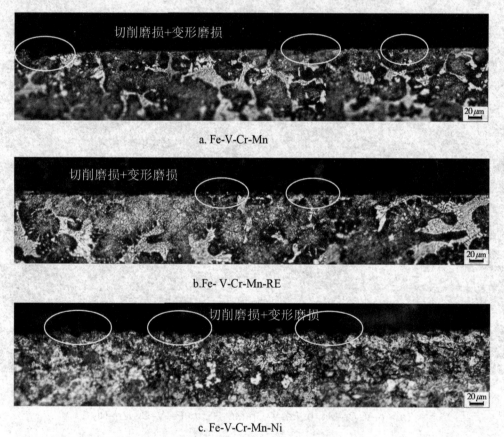

a. Fe-V-Cr-Mn

b.Fe- V-Cr-Mn-RE

c. Fe-V-Cr-Mn-Ni

Fig. 5-4-10   Cross sectional metallographic photos parallel wear direction (observation direction ①)

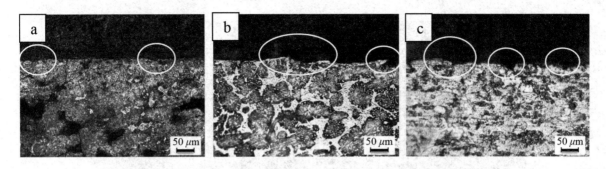

Fig. 5-4-11　Cross sectional metallographic photos perpendicular to
wear direction（observation direction ②）
（a: Fe-V-Cr-Mn, b: Fe-V-Cr-Mn-RE, c: Fe-V-Cr-Mn-Ni）

5.5　热处理工艺对新型Fe基Fe-V-Cr-Mn合金耐磨性能的影响

5.5.1　引言

　　热处理技术被广泛应用于机械制造的各个领域,可以用来提高材料的机械性能,消除残余应力和改善金属的切削加工性。热处理一般通过将金属材料放在一定的加热炉内加热、保温,然后在不同的冷却介质内或冷却方式下冷却,通过改变材料表面或内部的金相组织结构,使晶粒细化、组织均匀,提高材料的机械性能,消除残余应力,改善金属的切削加工性等。

　　本节主要对比Fe-V-Cr-Mn铸态下和经过正火后组织与性能的变化。

5.5.2　实验材料及金相组织

　　本章所用的材料是铸态下的Fe-V-Cr-Mn和正火后的Fe-V-Cr-Mn[以下用Fe-V-Cr-Mn（normalizing）表示],正火温度为1 200 ℃,保温时间为3个小时,从炉中取出后空冷至室温。

　　Fe-V-Cr-Mn与Fe-V-Cr-Mn（normalizing）的金相组织对比如Fig.5-5-1所示。从Fig.5-5-1中可以看出,合金经过正火处理后,组织中钒的碳化物发生了明显的球化现象,尺寸也有所增加,而Cr的碳化物较铸态下变得短小。Fig.5-5-2反映的是Fe-V-Cr-Mn（normalizing）中V, Cr元素的分布。正火后,V还是主要分布于它的碳化物中,而Cr在基体中的含量较铸态下有明显增加。对合金正火后的组织进行X射线衍射仪分析,如Fig.5-5-3所示,从中可以看到正火后合金的基体组织组成基本没有变化,依然为奥氏体+铁素体。

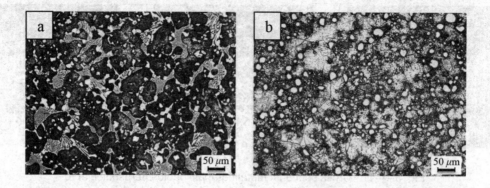

Fig. 5-5-1 Metallographic structures of as cast and normalizing Fe-V-Cr-Mn
（a: Fe-V-Cr-Mn, b: Fe-V-Cr-Mn（normalizing））

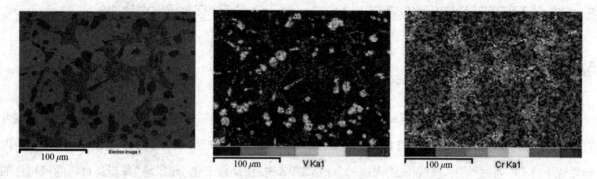

Fig.5-5-2 The distribution of V, Cr in Fe-V-Cr-Mn（normalizing）

84

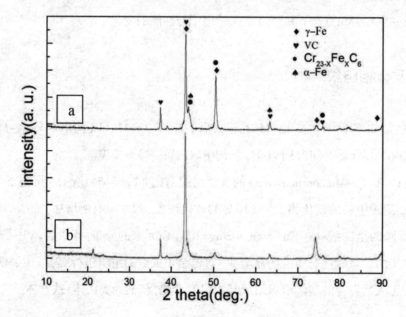

Fig.5-5-3 X-ray diffraction spectrum of Fe-V-Cr-Mn before and after normalizing
（a: Fe-V-Cr-Mn, b: Fe-V-Cr-Mn（normalizing））

5.5.3　耐磨性能及冲击韧性对比

5.5.3.1　磨粒磨损性能对比

用洛式硬度计测量合金正火后的表面硬度，结果如Table5-5-1所示。从表中可以看出，正火后合金的硬度比铸态下降了，这与正火后合金中铁素体含量增加有关。Fig.5-5-4为V-Cr-Mn合金铸态下与正火态下SUGA磨粒磨损实验结果对比。结果显示，正火后合金的磨损量比铸态下增加了12%，意味着经过正火热处理后，合金的耐磨性反而下降了。测量磨损前后的显微硬度可知，正火后的合金在磨损表面也发生了加工硬化现象，如Fig.5-5-5所示。

Table5-5-1　The hardness of Fe-V-Cr-Mn before and after normalizing（HRC）

材料	硬度值（HRC）					平均值
	1	2	3	4	5	
Fe-V-Cr-Mn	47.5	48	47.5	47.9	48	47.8
Fe-V-Cr-Mn（normalizing）	44	47.9	47.9	45.8	47.5	46.6

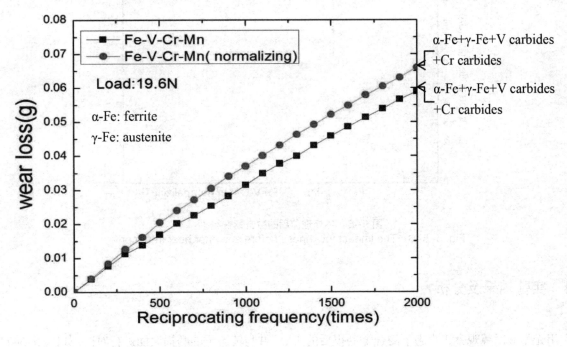

Fig. 5-5-4　Abrasive wear results of before and after normalizing

5.5.3.2　冲击韧性对比

本章对正火后的Fe-V-Cr-Mn做了冲击韧性实验，并将其与铸态下合金的冲击韧性进行对比，如Fig.5-5-6所示。冲击试样为10 mm×10 mm×55 mm无缺口的标准试样。结果显示，正火后合金的冲击韧性下降了20%左右。

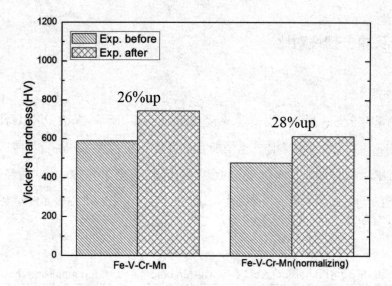

Fig. 5-5-5 Change of Vickers hardness for different materials after test

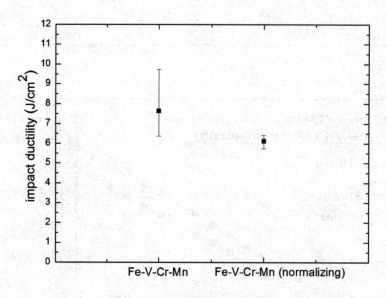

图 5-5   热处理前后的冲击韧性结果
Fig. 5-5-6 The impact toughness before and after heat treatment

5.5.4   实验结果及分析

应用光学显微镜观察正火态下的合金磨损后的表面，并与铸态下的磨损表面进行对比，如Fig.5-5-7为放大50倍下的磨损表面形貌。从图中可以看出，正火后的磨损表面较铸态下表现得更为平整。

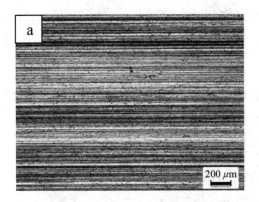 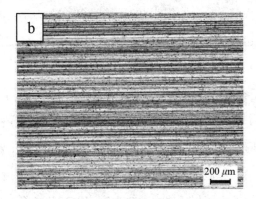

Fig. 5-5-7　Worn surface morphology of before and after heat treatment（50×）
（a：Fe-V-Cr-Mn, b：Fe-V-Cr-Mn（normalizing））

　　将磨损表面放大500倍后的表面形貌如Fig.5-5-8，从图中可以更加清晰地看到正火后的磨损表面划痕较铸态下少。同时在球状碳化物较多的区域，观察发现划痕深度更浅，也更为平整。

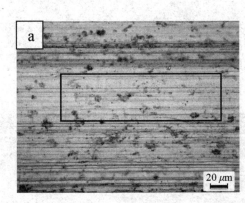 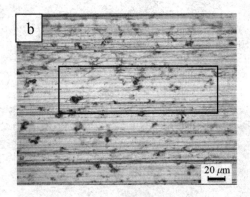

Fig.5-5-8　Worn surface morphology of before and after heat treatment（500×）
（a：Fe-V-Cr-Mn, b：Fe-V-Cr-Mn（normalizing））

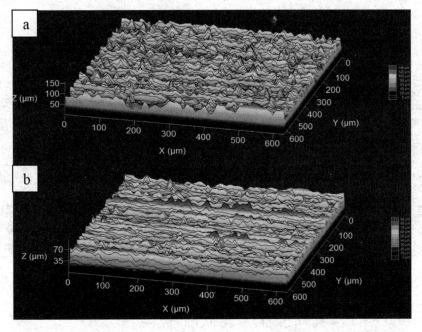

Fig. 5-5-9　Comparison of 3D images of worn surface before and after heat treatment
（a：Fe-V-Cr-Mn, b：Fe-V-Cr-Mn（normalizing））

从正火前后合金磨损表面的三维图像分析来看,如Fig.5-5-9所示,铸态下的磨损表面起伏在0~150 μm之间,起伏较大。正火态下的磨损表面起伏在0~70 μm之间,总体上较铸态下平整,但出现了连续的高峰,分析可能是由于正火后合金表面的硬度下降,基体较铸态下更容易剥落。

对正火后的材料同样做了平行于磨损方向及垂直于磨损方向的切割。Fig.5-5-10是平行于磨损表面切割后从方向①观察表面变化的金相照片(500×)。从图中可以看出,正火前后的材料磨损既产生了切削磨损,也产生了变形磨损。虽然正火后使VC的形状有所改善,减少了对基体的割裂作用,但由于正火后合金表面硬度的下降,使得磨粒磨损时基体容易被破坏,对碳化物的支撑作用变弱,整体表现为耐磨性下降。Fig.5-5-11是垂直于磨损表面切割后从方向②观察表面变化的金相照片(500×)。从图中可以看出,在平整区域内,正火后的磨损表面较铸态下更平整,这也进一步地证明了正火后的基体更容易整体剥落。

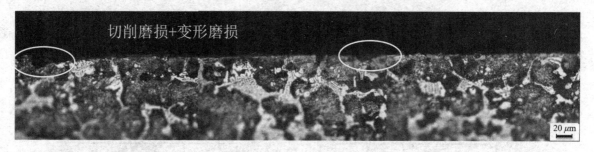

a.Fe-V-Cr-Mn

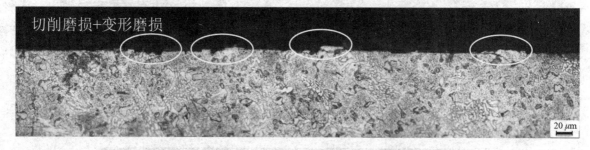

b. Fe-V-Cr-Mn (normalizing)

Fig. 5-5-10　Cross sectional metallographic photos parallel wear direction (observation direction ①)

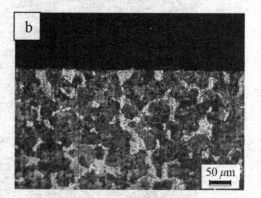

Fig. 5-5-11　Cross sectional metallographic photos perpendicular to
wear direction (observation direction ②)
(a: Fe-V-Cr-Mn, b: Fe-V-Cr-Mn (normalizing))

5.6　结论

本课题为了提高现有的以高铬铸铁、高锰钢为主要材质的球磨机衬板和磨球的使用寿命,开发出一种新型Fe基 Fe-V-Cr-Mn合金。新合金较高铬铸铁及高锰钢在磨粒磨损条件下表现出更好的耐磨性能。使用新型Fe基Fe-V-Cr-Mn合金制造的球磨机衬板及磨球的使用寿命会更长,成本也相应会降低很多。

本文主要研究内容是通过对比三组不同的材料:①Fe-V-Cr-Mn, Mn13Cr2, Cr27Mo1;②Fe-V-Cr-Mn, Fe-V-Cr-Mn-RE, Fe-V-Cr-Mn-Ni;③Fe-V-Cr-Mn, Fe-V-Cr-Mn(normalizing)。通过分析它们的金相组织、硬度、耐磨性、冲击韧性、磨损表面形貌的差异,得出以下结论:

1. Fe-V-Cr-Mn的金相组织为α-Fe+γ-Fe+V的碳化物+Cr 的碳化物。

2. Fe-V-Cr-Mn中碳化物所占面积比例为20%~25%。

3. Fe-V-Cr-Mn的耐磨性是高铬铸铁的2倍,是高锰钢的2.5倍,冲击韧性略低于高铬铸铁。

4. Fe-V-Cr-Mn加入0.2%的稀土后,合金中VC的尺寸明显变小,碳化物所占面积比例减少到15%~18%,不利于材料的抗磨粒磨损性能,耐磨性下降了15%,冲击韧性也略有下降。加入稀土并没有起到改善合金组织与性能的作用。

5. Fe-V-Cr-Mn加入3%的Ni后,合金表面硬度下降了15%,耐磨性下降了31%,冲击韧性也略有下降。

6. 正火后虽然使合金中碳化物的形貌得到了改善,但导致合金整体硬度下降,耐磨性降低了12%,冲击韧性降低了20%。

通过本课题的研究发现,新型Fe基Fe-V-Cr-Mn合金在铸态下就已经表现出良好的耐磨性能和抗冲击性能,不需要经过一般的热处理,就可以用来代替现有的以高铬铸铁及高锰钢为材质的球磨机衬板和磨球,满足它们在磨粒磨损工况条件下的实际要求。同时,所设计的新材料也不需要添加稀土和Ni进行改性处理,大大减少了制造成本。在以后的研究中,可以通过设计更合理的化学成分配比,以及使用其他的热处理工艺,使新材料的力学性能进一步提高,成本继续降低。

参考文献

[1] 徐流杰, 韩世忠, 韩明儒, et al. 高钒钢的组织与性能 [M]. 北京: 科学出版社, 2010: 58–61.

[2] 王铀, 徐永利, 扈延光, et al. 关于磨损评价的评价; proceedings of the 第二届全国青年摩擦学学术会议, 中国陕西西安, F, 1993 [C].

[3] 陈华辉. 耐磨材料应用手册 [M]. 北京: 机械工业出版社, 2012.

[4] 赫罗绍夫著, 胡绍衣译. 金属的磨损 [M]. 北京: 机械工业出版社, 1966.

[5] 克拉盖尔斯基著, 汪一麟译. 摩擦磨损计算原理 [M]. 北京: 机械工业出版社, 1982.

[6] YAER X, SHIMIZU K, MATSUMOTO H, et al. Erosive wear characteristics of spheroidal carbides cast iron [J]. Wear, 264 (11–12), 2008: 947–57.

[7] 翟启杰. 钒在铸铁中的作用及含钒铸铁 铸铁中的微量元素讲座之四 [J]. 现代铸铁, (04), 2001: 25-8.

[8] 冯春雨. 浅谈合金元素在钢中的作用 [J]. 冶金标准化与质量, (08), 1996: 20-3.

[9] 王建华, 任立军. 高锰钢加工硬化机理研究 [J]. 煤矿机械, (01), 2003: 24-7.

[10] 杨芳, 丁志敏. 耐磨高锰钢的发展现状 [J]. 机车车辆工艺, (06), 2006: 6-9.

[11] 李卫. 我国耐磨材料耐磨铸件的标准化; proceedings of the 21世纪全国耐磨材料大会——第九届全国耐磨材料磨损失效分析与抗磨技术学术会议, 中国北京, F, 2000 [C].

[12] 李海鹏, 梁春永, 王立辉, et al. 铬系白口铸铁的研究进展 [J]. 中国铸造装备与技术, (05), 2006: 8-12.

[13] 孙凯, 邢兆林, 付拴拴, et al. 高铬铸铁热处理工艺研究现状 [J]. 热加工工艺, (14), 2012: 178-9.

[14] 褚祥治, 张晓娟, 苑少强, et al. 高铬铸铁的研究应用与发展 [J]. 唐山学院学报, (03), 2011: 23-5.

第6章 表面改性材料冲蚀磨损特性

6.1 序言

截至上一章,已评估各种球状碳化物铸铁的耐冲蚀磨损性能。其结果是,该材料由于组织中含有球状VC碳化物而获得优异的抗冲蚀磨损性能。其特点是,基体组织为软相,VC的硬度很高,以及VC和基体具有非常良好的接合力。这种结构有可能抑制冲蚀颗粒的碰撞造成的磨损。此外,由于VC是球形的,当发生碰撞时,应力集中被分散,能量被吸收掉,结果呈现出优异的抗冲蚀磨损性能性。然而,这些材料还是有以下几个缺点,有必要进一步完善:

1. 球状碳化物铸铁在生产过程中,VC的球化是在高温下进行的,其温度控置需要较高的技术。

2. 保证不了VC能够均匀地分散在基体组织中。

3. 球状VC碳化物的直径为10~20 μm,与冲蚀颗粒的平均粒径408 μm相比小得多,所以当受到冲蚀颗粒的冲击时,容易被基体一同去除的可能性很大,因此球状碳化物的抗磨损效果受到一定影响。

所以,本章研究中,我们以表面改性为目标,在各种材料的表面层镶嵌球状的大颗粒碳化钨(WC),通过利用铸渗工艺开发出表面增强耐磨复合材料。

6.2 铸渗法(cast-in insertion)的介绍

首先,介绍常见的表面改性方法:

1. CVD(化学气相沉积);

2. PVC(物理气相沉积);

3. 喷涂;

4. 电镀;

5. 渗碳;

6. 渗氮;

7. 堆焊;

8. 铸渗

等等。

然而,上述方法中1~6的表面改性方法具有制造成本普遍高的缺点,而且改性层的厚度很薄,只有数十微米，不能适用于冲蚀磨损环境。堆焊工艺可以得到较厚的表面堆焊层（2~5 mm），但是得到均匀大面积的堆焊层比较困难，加之也不能被应用到较小的零部件中。[1-4]

然而，铸渗方法是指固体金属和熔融金属通过铸造方法接合的技术。这种方法是，首先把固体金属安放在铸型的特定部位，然后把熔融金属注入铸型的一种加工方法。该方法成本低，尺寸和形状相对自由，对铸造材料没有限制，设计变更也很容易，甚至对表面增强层的厚度设计自由度也较大。

本章中使用砂型铸造方法，制备试验样品。在Fig.6-2-1中表示的是该铸造工艺方案的模型。在砂型中，首先放置孔径小于碳化钨（WC）颗粒的铁丝网，然后放置球形碳化钨（WC）颗粒在规定孔径上，最后用同样孔径的铁丝网盖住，防止液态金属注入型腔时硬质碳化钨的移动。增强颗粒是平均粒径为6 mm，维氏硬度为1 500 HV球形的碳化钨（WC）。选定上述碳化钨（WC）的理由是，因为其硬度高，常被用来当作抵抗磨粒磨损的耐磨材料。此外，选择大颗粒尺寸，以便尽可能地抑制铁水的流动。球形形状的选择是因为球状碳化物铸铁中析出的球状VC分散应力而得到良好的效果。

通过铸渗法制造的铸锭表面和横截面的外观照片如Fig.6-2-2所示。（A），（B）是表面的外观，（C），（D）是相对于表面的截面状态。从表面照片可以看出，灰铸铁（FC）和含碳化钒的球状碳化物铸铁（Spheroidal Carbides cast Iron: SCI-VCrNi）材料表面被铁水充满，证实充型健全。从照片（C），（D）可以看出，碳化钨（WC）粒子的基体材料表面无移动均匀的分布。最后，母材和表面改性材料加工成50 mm×50 mm×10 mm试验片，进行了冲蚀磨损试验。另外，把表面改性材料的增强体与母材的接合处部位进行抛光，用光学显微镜观察。

6.3　试验方法及实验结果

在冲蚀磨损试验中，与前面的章节一样，采用了吸引式喷砂机。使用的冲蚀颗粒为，无规则的平均粒径为410 μm，硬度为1 030 HV的石英砂。在实验过程中，冲蚀粒子本身也被磨损，其尺寸发生改变，所以根据要求每个实验完成后要更换冲蚀颗粒砂子。靶材为已制备的铸渗材料，冲蚀速度为100 m/s，试验冲蚀角度分别为30°，60°，90°，喷砂量为5~7 g/s，试验时间定为3 600 s。通过灵敏度为0.1 mg电子天平来称量试验前后的重量，最后计算出试验中的磨损量。

Fig.6-3-1表示FC母材和FC基铸渗材料经过3 600 s冲蚀磨损后的损伤速度和冲蚀角度之间的关系。FC基材表现出在低角度30°侧，损伤速度变为最大，随着冲蚀角的增大，损伤速度逐渐减小，在60°取最小值。此后损伤速度随冲蚀角度继续增加而增加，在90°时取另一个峰值的两个峰的曲线。然而，FC基铸渗材料在所有的角度都能够控制磨损。

结果是，在其损伤速度在30°时减少了50%，在60°时约减少了22%，在90°时约减少了31%。虽然FC基材在低角度时由于切削磨损呈现峰值，但通过铸渗法连接碳化钨（WC）时角度依赖性得到了改善。基于上述结果，本身的耐冲蚀磨损性能优异的SCI-VCrNi的表面也通过铸渗法结合碳化钨（WC），其实验结果在Fig.6-3-2中。从图中可以看出，虽然整体的抑制磨损情况并不像FC那样明显，但在90°角度时效果明显，磨损量相对于母材被抑制约为22%。可以认为，通过应用碳化钨（WC）铸渗到母材的方法可以明显提高靶材的抗冲蚀磨损能力。为了揭示其原因,研究碳化钨(WC)本身的特点及碳化钨(WC)与基材间的接合力,也就是研究碳化钨(WC)与母材之间形成的反应层的性能。

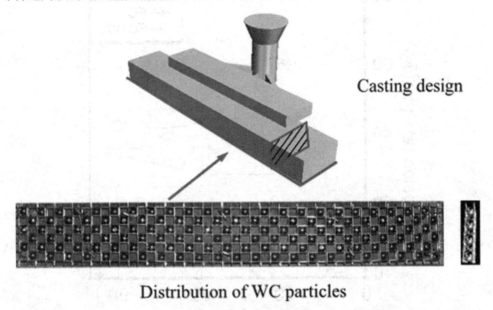

Fig. 6-2-1 Casting design and the distribution of WC particles on the surface in cast-in insertion process

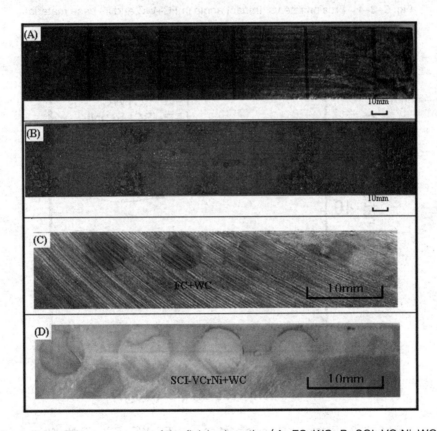

Fig. 6-2-2 The appearance of the finished casting(A: FC+WC, B: SCI-VCrNi+WC)

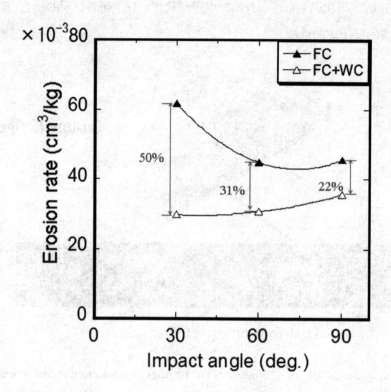

Fig. 6-3-1   Erosion rate vs. Impact angle in FC+WC and its base material

94

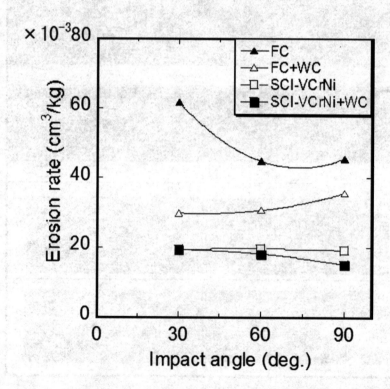

Fig.6-3-2   Erosion rate vs. Impact angle in FC+WC, SCI-VCrNi+WC
and their base materials

6.4　讨论

6.4.1　磨损面的宏观观察

针对各种铸渗材料的磨损面,为确定碳化钨(WC)是否有效地接合于母材,把试验后的磨损面利用数字机照相观察它们的表面状态。在前几章的研究[1, 4-7]中已看到,不管什么样的材料,低角度的受损区域宽,朝向冲蚀方向伸长。在高角度变得接近于圆形时,损伤面积较小。但是,从本次试验中铸渗材料的宏观照片上看,如Fig.6-4-1所示,可以看出在所有磨损表面都已经露出了碳化钨(WC),而且碳化钨(WC)的磨损特别小,而其周围的基体被磨损掉,只有碳化钨(WC)主体的部分抑制了磨损。在FC基铸渗材料中,在30°和90°冲蚀角度时的试样中碳化钨(WC)均匀地布置在其表面,并且很好地控制着冲蚀磨损。在60°情况下的试样中,虽然碳化钨(WC)稍微从原来的位置偏移,在磨损表面出现稍有排列不均匀的情况,但冲蚀磨损也得到了控制。在SCI-VCrNi基铸渗材料中可以看出,虽然围绕碳化钨(WC)的基体组织在被磨掉,但相比FC基铸渗材料,磨损量甚少,在每个冲蚀角度的磨损量也非常小。

6.4.2　对各试样的反应层的微观观察

为了阐明上述铸渗材料的碳化钨(WC)在表面接合程度的好坏,从试样的接合部位取样、研磨、抛光,进行光学显微镜观察。还有,利用显微维氏硬度计测定了反应层的硬度。结果显示,在Fig.6-4-2和Fig.6-4-3中,在FC基铸渗材料中的碳化钨(WC)和FC基体之间存在反应层,其厚度约为170 μm,维氏硬度大概为1 083 HV。在SCI-VCrNi基铸渗材料中,也形成类似的反应层,但其厚度为仅为5 μm,维氏硬度为928 HV。由此可以判断,耐磨损性能的提高与形成反应层有关。

6.4.3　反应层的 EPMA 面分析结果

为了检验各铸渗材料反应层的成分,利用EPMA进行了面分析。分别在Fig.6-4-4中表示FC基铸渗材料和Fig.6-4-5中表示SCI-VCrNi基铸渗材料反应层的EPMA面分析结果。可以看出,在FC基铸渗材料的反应层中,主要成分为W,还有少量Fe, Co, C等。此外还可以确认,Co有向基体金属扩散的倾向。在SCI-VCrNi基铸渗材料的反应层中能够确认,包括主要成分由W与少量的Fe, Cr组成。由于W含量高反应层的形成,其硬度显著提高,结果被认为具有了良好的冲蚀磨损性能。可以认为,形成在碳化钨(WC)和母材的边界处的反应层的厚度,尽管随基体材料的结构不同而变化,但只要能够形成高硬度的反应层,材料表面和碳化钨(WC)之间的接合力增加的结果就能提高耐冲蚀磨损性能,这与反应层的厚度薄厚没有多大关系。

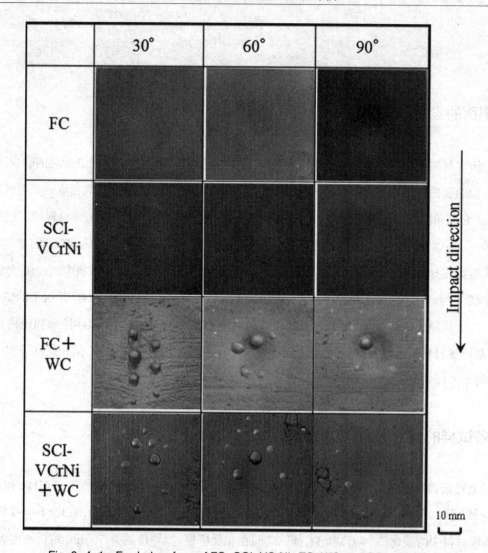

Fig. 6-4-1   Eroded surface of FC, SCI–VCrNi, FC+WC and SCI–VCrNi+WC

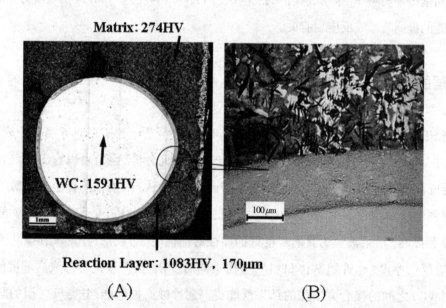

(A)   (B)

Fig. 6-4-2   Wideness and Hardness of Bonded Area in FC+WC

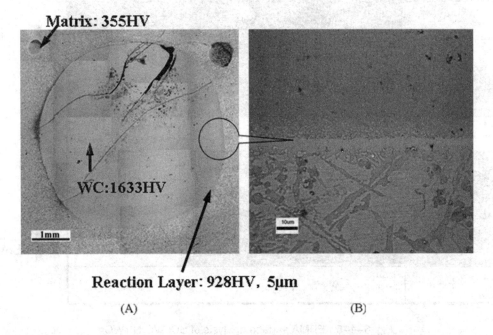

Fig. 6-4-3　Wideness and Hardness of Bonded Area in SCI-VCrNi+WC

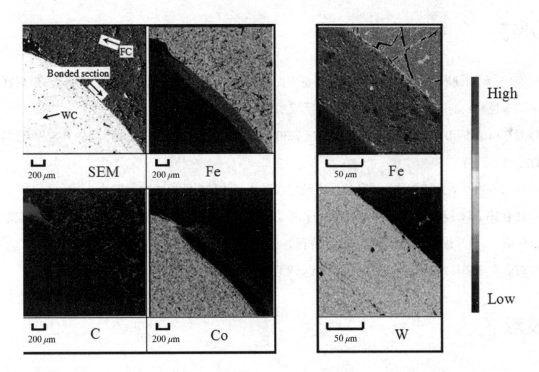

Fig. 6-4-4　EPMA surface analysis of FC+WC

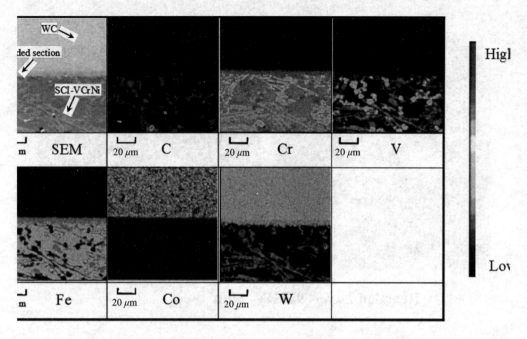

Fig. 6-4-5　EPMA surface analysis of SCl-VCrNi+WC

6.5　小结

本研究中,利用硬质碳化钨(WC)颗粒,制备表面增强材料,并对其进行了冲蚀磨损实验。然后评估和解析这些表面改性材料的冲蚀磨损特性和形成的反应层。总结结论如下:

(1)对于FC基铸渗材料,损伤速度在30°减少50%,在60°约减少22%,在90°减少31%,而且其冲蚀角度依赖性也得到改善。

(2)通过在磨损表面镶嵌硬质碳化物的方法可以提高基体材料的抗冲蚀磨损性。

(3)碳化钨(WC)和母材边界形成的反应层厚度和硬度,随母材的性质变化而变化,即使反应层较薄,一旦形成反应层,抗冲蚀磨损特性就可以得到改善。

(4)对表面增强材料制备过程中球形碳化钨(WC)的位置固定需要进一步探讨。

参考文献

[1] SHIMIZU K, NOGUCHI T, KAMADA T, et al. Progress of erosive wear in spheroidal graphite cast iron [J]. Wear, 198(1), 1996: 150-5.

[2] 麻生節, 後藤正, 池浩, et al. 27%クロム白鋳鉄による超硬合金の鋳ぐるみと組織評価 [J]. 鋳造工学, 70(12), 1998: 878-83.

[3] 麻生節, 中西誠, 後藤正, et al. WC粉末による鋳鉄の表面硬化 [J]. 鋳造工学, 73(3), 2001: 155-60.

[4] 新巴雅尔, 清水一, 桃野正, et al. 高マンガン球状炭化物鋳鉄のサンドエロージョン摩耗特性 [J]. 鋳造工学, 78(10), 2006: 510-5.

[5] K.SHIMIZU T N A S D. Trans of AFS, 101 (1993) : 225-9.

[6] KAZUMICHI SHIMIZU, NOGUCHI T. IMONO, 66 (1994) : 7.

[7] SHIMIZU K, NOGUCHI T. Erosion characteristics of ductile iron with various matrix structures [J]. Wear, 176 (2) , 1994: 255-60.

第7章　新型柱状碳化钨表面增强耐磨复合材料的磨粒磨损性能

7.1　硬质合金增强钢铁基耐磨复合材料的研究现状

　　硬质合金增强钢铁基耐磨材料是表面增强双金属耐磨复合材料的重要分支。硬质合金有很多优异的性能，尤其是由于它的高硬度和高耐磨性，被广泛应用在矿山开采、冶金熔炼、机械制造等领域。[1]但是，由于它的硬度高、脆性大且价格昂贵等，在一些承受强冲击的易损件上使用或单独使用时，极易发生破碎或脆断而造成零部件失效，使其应用受到了一定程度的限制。所以，从降低成本，提高工作效能和使用寿命考虑，近些年硬质合金增强钢铁基耐磨复合材料得到了重视，成为新的开发热点领域之一。[2]

　　硬质合金增强钢铁基耐磨复合材料是由两种不同性能的材料结合为一体组成的新材料，通常包括基体和增强相两部分，其中增强相以自身的高耐磨性保护基体，基体又把硬质合金结合在一起并支撑硬质合金，在二者相互协同作用下，抵抗磨损，主要用于耐磨、耐热及耐蚀等复杂工况。

7.1.1　基体和硬质合金的选择

　　由于复合材料中基体对硬质合金起支撑的作用，所以必须有一定的硬度和强度。因此，通常由实际服役条件决定选用的基体，如普通碳素钢、高锰钢、合金钢和球墨铸铁等，但这些基体硬度偏低，在磨损过程中，磨损较快，失效早，导致硬质合金的高耐磨性能无法完全发挥或碳化钨（WC）硬质相剥落，造成投资大，收益小。等温淬火球墨铸铁（Austempered ductile cast iron：简称ADI）是近30年发展起来的一种高强度、高性能的新一代球墨铸铁材料，在各类不同要求的耐磨性应用中被广泛使用，取得了显著的经济效益。它有高的强韧性，抗磨损，重量轻，好的吸震性，成本低，能源损耗少，有应变强化能力，因而具有优越的耐磨性。[3]如果用作硬质合金增强复合材料的母材，是非常有潜力的。

　　硬质合金是由高熔点、高硬度的碳化物和黏结金属制成的高硬度材料，其中常见的碳化物主要有碳化钛、碳化钨、碳化钒、碳化铌等。[4]黏结相赋予硬质合金一定的硬度和韧性，而碳化物相保证其高的硬度和抗磨性。WC-Co系硬质合金，由于其具有许多优异的性能[5]，目前产量最大、适用面最广。硬质合金的化学性能稳定，热膨胀系数较小，弹性模量、冲击韧性、抗弯强度和抗压强度高。特别是由于它的高硬度与优异的耐磨性，用其制作的耐磨件使用寿命远高于其他金属耐磨件的寿命，但WC-Co系硬质合金，同样存在硬度高、脆性大的问题, 因此将利用高硬度、高耐磨性的碳化钨硬质合金和强度高、韧性好且抗磨损的奥贝球铁, 制备碳化钨表面增强耐磨复合材料, 使碳化钨与奥贝球铁发挥各自的优点, 最终获得既抗磨损又抗冲击的复合材料是较为可行的研究思路。

7.1.2　硬质合金增强钢铁基耐磨复合材料的制备方法

关于硬质合金增强双金属抗磨材料的研究, 西安建筑科技大学的王双成等人[4]在钢铁基双金属复合材料研究的基础上, 采用电磁感应熔渗法制备了硬质合金增强的钢基耐磨材料。研究表明, 基体与硬质合金冶金结合良好 (且硬度过渡均匀), 复合材料的相对耐磨性随着载荷的增加而明显提高, 当达到5.25 kg时, 其耐磨性为45钢的近5倍。[6]西安交通大学的鲍崇高等人[7]通过消失模铸造技术, 获得了钴基硬质合金局域化增强高铬铸铁基表层复合材料, 研究结果表明, 增强体与基体界面结合良好, 其相对耐磨性是基体材料的8.24倍。刘亚民等利用复合铸造技术制备了硬质合金—球墨铸铁复合材料, 发现采用合适的工艺, 可获得牢固结合的复合材料, 而硬质合金的组织、性能基本不受影响。[8]

目前, 硬质合金增强钢铁基复合材料的制备方法通常有复合浇铸、复合堆焊、镶铸、焊接镶嵌等复合或组合工艺方法。以上制备方法中钎焊、堆焊等复合方法, 在实际生产中需大量的专业设备, 成本高, 生产效率低, 还有不易控制堆焊层或钎缝不致密等缺陷, 因此很难规模化推广应用。[9]其中镶铸法是将预制成一定形状的抗磨镶件置入型腔, 随后往型腔浇入韧性好的母材 (基体), 在熔融母材的热作用下, 基体与抗磨镶件在界面上实现良好的冶金结合, 使镶块牢固地镶嵌在母材中, 最终获得抗磨复合材料。铸造镶铸法优点在于: 耐磨镶块可按工件具体使用要求选用不同材质和形状的预制块, 镶铸件的表面质量相对容易控制, 工艺简单, 在普通的铸造条件下就能生产。镶铸法制备的抗磨件现已获得成功, 用其制造的导卫、铲齿、衬板、磨头板等已获得成功, 并用于生产, 其抗磨性能好, 经济效益显著。[10, 11]

7.1.3　本课题研究的目的及意义

综上所述, 本课题研究的目的是将通过结合消失模铸造与镶铸法开发廉价的钢铁基复合材料, 通过该工艺将具有高硬度、高耐磨性的碳化钨增强体与综合性能良好的奥贝球铁复合在一起, 制备工作部位高硬度、基体部位高韧性和高强度的铁基复合材料, 利用碳化钨自身的强度和硬度, 保护基体不被磨损, 同时基体起着把碳化钨结合在一起并支撑碳化钨的作用, 二者共同抵抗磨损, 实现耐磨部件较高的耐磨性, 以满足服役要求。

这种复合材料增强体和基体之间达到良好的冶金结合, 结合强度高, 在增强体保护基体材料和基体材料有效支撑增强体的协同作用下, 达到既硬又韧的效果, 增强体不易脱落, 获得性能良好的复合材料。

随着工业的快速发展, 破碎机被广泛用于矿山、建筑等领域, 用来破碎不同的物料。其中对破碎机衬板、颚板和锤头等的需求量最多, 同时也是消耗量最大的。而现有的衬板等通常都是高锰钢、高铬铸铁等常规的耐磨材料, 虽然可以在一定程度上满足破碎机的使用要求, 但是由于材料本身的特点, 其应用受限。因此, 研制出一种表面强化铁基耐磨复合材料,使其耐磨性比单质基体好2~3倍,在复杂的工况条件下使用寿命更长,具有优异的耐磨性是非常有意义的。

7.2 柱状碳化钨表面增强耐磨复合材料的制备

破碎机锤头、挖掘机截齿等零部件在外力的作用下直接与岩石、砂土等物料接触,引起严重磨损并消耗大量钢铁耐磨件。因此,易损件需要表面改性来提高材料的表面性能,基于这种原因开发了表面增强耐磨复合材料。

本课题首先通过铸造方法制备出复合材料并进行等温淬火处理,然后对复合材料进行磨粒磨损实验等相关实验,最后分别与等温淬火前后的基体材料、等温淬火前的复合材料进行相应的对比,并分析其中原因。具体步骤为:首先通过真空消失模工艺制备球铁和柱状碳化钨表面增强球铁基复合材料,然后对其进行等温淬火处理,制备出奥贝球铁及奥贝球铁基复合材料,最后分别对球铁、奥贝球铁、球铁基复合材料及奥贝球铁基复合材料进行耐磨性能对比与分析。

7.2.1 复合材料增强体与基体的选择

7.2.1.1 增强体材料

硬质合金是采用粉末冶金方法,以高熔点、高硬度的碳化物和黏结金属制成的高硬度材料,其中碳化钨硬质合金应用最为广泛,但是存在用其制作的耐磨件硬度高、脆性大,使用在工况复杂的情况时容易产生脆断,且实际生产中成本太高等问题。[7]因此,近年来硬质合金增强钢铁基复合材料得到了广泛的应用。所以,本文以碳化钨硬质合金为增强体,来制备工作部位高硬度、基体部位有较好强韧性的碳化钨表面增强铁基耐磨复合材料(抗磨又抗冲击),可以降低成本,节约大量的碳化钨硬质合金。

本研究复合材料复合层中增强体材料,选用以钴为黏结剂的碳化钨硬质合金,其主要成分为92wt.%碳化钨(WC),8wt.%Co。把直径为6 mm的碳化钨棒用线切割成长度10 mm的短棒,其密度为14.5~14.9 g/cm³,硬度为89 HRA,抗拉强度≥1 500 Mpa,其较高的强度和硬度对材料耐磨性能的发挥具有重要影响。

7.2.1.2 基体材料

等温淬火球墨铸铁,也称奥贝球铁(简称ADI)。研究表明,这种等温淬火组织是由高碳奥氏体(25%~40%)和针状铁素体组织构成,不含有渗碳体,是一定成分的球墨铸铁经等温淬火后得到的铸铁材料,是材料领域新科技的产物,具有优良的综合性能,是值得推广的优良工程材料。[2]

由于奥贝球铁中残余奥氏体的存在,当受到强力的摩擦磨损时,除了会产生加工硬化外,还因为在摩擦表面会产生应力或应变诱发马氏体相变,伴随马氏体的转变将产生约4%的体积膨胀,因而使摩擦表面形成压应力,从而提高工件的疲劳磨损抗力。正因为奥贝球铁具有这种特殊的性能,目前主要应用于耐磨、耐冲击和有高疲劳强度要求的零部件上,如矿山机械、电力、冶金等行业。[2]

因此,如果用(WC)–Co硬质合金作为复合材料的增强相,对工作面进行有效的表面强化,以奥贝球铁作为复合材料的基体,并且两者形成良好的冶金结合时,这种复合材料就同时拥有了WC-Co硬质合金的高硬度及高耐磨性和奥贝球铁的强韧性,并且可以降低使用昂贵的硬质合金带来的成本。

7.2.2　复合材料的制备工艺

消失模铸造技术是用泡沫塑料(EPS, EPMMA)制成与零件结构、尺寸完全相同的实型模具,再涂覆具有强化和光洁作用的耐火涂料,烘干后埋箱造型并负压浇注熔融的金属液,最后泡沫模具被金属取代,成为精确成型的铸造工艺。与传统铸造技术相比,有很多独特的优越性。[12]

1. 铸造质量方面:铸件尺寸精度高,均匀一致;铸件表面质量好,降低或排除了气孔、缩孔等缺陷。

2. 适应范围广:适合复杂铸件的生产,可适用于大批量的生产。

3. 投资少、见效快:工厂的设计灵活,占地面积小,对工人的技能要求低且用人少;不需要大量的清理设备;无铸件飞边毛刺。

4. 设计灵活:传统铸造要考虑很多的工艺因素,而消失模铸造大大减少了铸造工艺限制,可实现最优化的设计。

5. 环境保护:排放污染少,简化了环保措施。

与普通铸造相比,不用取出泡沫塑料模,所以不需要设计拔模斜度,对工艺设计带来很多便利。所以本课题将在已有研究的基础上,结合真空消失模铸造工艺制备出碳化钨表面增强铁基复合材料。

7.2.3　基体的制备过程

7.2.3.1　化学成分的确定

获得良好的组织状态和高性能的基本条件是选择适当的化学成分,化学成分及一些合金元素对ADI基体组织及性能的影响如下:

(1)碳和硅　碳高时易使石墨漂浮,使铸件中夹杂物的数量增多,影响性能;碳低时易产生缩松和裂纹等缺陷,控制在3.4%~3.8%较为合适。硅是强石墨化元素,能阻止渗碳体的析出,但硅高容易形成异形石墨及提高低温脆性,因此应控制在1.4%~2.4%。

(2)锰　锰有稳定和强化基体的作用,使碳曲线右移,有利于提高淬透性、强度及耐磨性。但锰易偏析,促进碳化物形成,又可阻碍石墨化,使石墨分布不均匀。过高的锰含量会降低ADI的强度和塑性。因此锰含量不宜过高,应控制在0.4%以下。

(3)磷和硫　磷和硫都是有害元素,尤其是磷有严重的偏析倾向,易在晶界处形成磷共晶,提高低温脆性,严重降低低温韧性,因此越低越好。硫可消耗球化剂,造成球化不稳定。为了保证良好的球化,以防过多的夹杂物产生和球化衰退,需要限制硫含量。

(4)铜、钼、镍　钼使碳曲线右移,能使铸件有良好的淬透性且能减少处理的时间。但较高的含钼量将降低ADI的塑性和强度。镍和铜都使碳曲线右移,同样提高淬透性。铜还能细化石墨和贝氏体针,使组织更均匀,有固溶强化的作用, 从而提高铸件的韧性和耐磨性。[13]

在化学成分的确定上, 综合考虑以上元素对ADI基体组织及性能的影响, 基体的化学成分设定如Table7-2-1所示。碳化钨硬质合金的化学成分见Table7-2-2。

Table7-2-1　Chemical composition of SGI　　　　　　　　　　　(wt%)

C	Si	Mn	Cu	Ni	Mo
3.6	2.5	0.3	0.7	1.2	0.3

Table7-2-2　Chemical composition of tungsten carbide　　　　　(wt%)

WC	Co
92	8

7.2.3.2　实验原材料

使用的原材料主要为硅铁、高碳锰铁、生铁、钼铁、废钢、纯镍板及纯铜块等。其主要成分为Table7-2-3中所示。

Table7-2-3　Chemical composition of raw materials　　　　　　　(wt%)

材料	C	Si	Mn	Al	Mo	Cu	S	P
硅铁	0.150	72.100	—	1.490	—	—	0.016	0.034
高碳锰铁	6.500	0.800	65.100	—	—	—	0.020	0.170
生铁	4.370	0.680	0.120	—	—	—	0.026	0.050
钼铁	0.050	0.290	—	—	57.600	0.350	0.095	0.030
废钢	0.200	0.400	0.450	—	—	—	0.100	0.040

7.2.3.3　熔炼及浇注

本实验利用中频感应电炉(额定容量30 kg, 60 kW的额定功率, 650 V的输出电压, 输出频率: 2 500 Hz。)对原材料如硅铁、高碳锰铁、生铁、钼铁、废钢、纯镍板及纯铜块进行熔炼及制备。铸造方法为消失模铸造, 其中制备实型模样时先切割与铸件形状相同的泡沫模样, 然后在模样表面涂上一层耐火涂料(如果涂料层太厚, 泡沫汽化时无法通过涂料层排出, 而太薄的话又起不到有效的支撑作用, 容易塌箱, 因此涂料层厚度一般为1~2 mm), 待模样干燥后再进行埋箱造型, 最后负压浇注。其中负压浇注时, 先将涂有耐火涂料的铸件模型放到可以抽真空的沙箱里(埋在干砂中), 振动紧实造型, 在抽负压下进行浇注(0.04~0.06 Mpa), 最后浇注完, 使金属液占据原来泡沫塑料模的位置, 在随后的凝固过程结束后获得所需的铸件。熔炼温度控制在1 520~1 550 ℃, 采用稀土镁合金球化剂进行球化处理, 用75硅铁进行孕育处理, 球化和孕育处理都将使用冲入法, 球化温度为1 470~1 500 ℃, 浇注温度为1 380~1 400 ℃。出炉前用铝丝脱氧。

7.2.3.4　奥贝球铁的制备工艺

为了解决球墨铸铁获得较高强度和硬度后能有良好的韧塑性与之匹配, 人们开发了铸件的奥氏体等温淬火技术。本课题研制的耐磨复合材料主要应用在采煤机截齿、挖掘机斗齿等工作在地质结构复杂的矿山易磨损件上，这些易磨损件齿尖部分需要有很高的耐磨性，而齿体又需要较强的抗冲击性能，所以对复合材料的球铁基体进行等温淬火处理，使基体处理成高强度、高韧性的奥贝球铁，以达到其理想的使用效果。

等温淬火球铁热处理共有3个阶段，要根据力学性能使用要求，工件的形状、尺寸及化学成分来制定淬火工艺参数[2]和参数组合。

（a）在850~950 ℃保温，使基体全部转变为碳饱和奥氏体。

（b）迅速淬入奥氏体等温转变温度260~380 ℃，使其避开珠光体转变而向铁素体转变。

（c）在等温转变温度范围内保温一定时间，获得针状铁素体+高碳奥氏体（碳含量1.8%~2.2%）组织。

（1）奥氏体化加热时间：奥氏体化温度越高，珠光体越多，硅越少则转变越快。综合考虑奥氏体内碳含量和均匀程度、合金元素的含量、试件的壁厚，在60~120 min范围内选择。

（2）等温淬火时间和奥氏体含碳量：等温淬火时间的长短在生产等温淬火球铁过程中起着主要作用。经过较短孕育期，铁素体在奥氏体中形成并生长。随着铁素体的生长，所排出的碳进入周围的奥氏体中，使奥氏体的含碳量增至1.2%~1.6%。这种奥氏体在室温时是稳定的，但力学上不稳定，还要在等温液中继续保温；当奥氏体碳含量增至1.8%~2.2%时，这种碳含量使奥氏体不但在热力学上稳定，力学上也稳定，这才是等淬球铁所应有的组织。而等温淬火时间太长时，高碳奥氏体将转变为更加稳定的铁素体和碳化物。应该尽量避免对力学性能有害的碳化物的出现。因此，一般等温淬火时间在45~120 min之间选择。

（3）等温淬火温度：经奥氏体化加热和保温后，将工件迅速淬入奥氏体等温转变温度的盐浴中进行等温转变。一般等温淬火温度为250~380 ℃。工程上可以通过调整奥氏体化温度、等温淬火温度及保温时间来达到所要求的性能组合。[13]

本实验要求淬火介质工作温度为320~380 ℃，因此选择了熔点略低于225 ℃的KNO_3和$NaNO_3$之比为1∶1的混合盐，其熔点为230 ℃，使用温度为300~500 ℃，对球铁进行等淬处理。

最后制定了等温淬火工艺参数：首先对制备好的球铁进行奥氏体化处理，温度为870 ℃，保温90 min，然后将迅速淬入370 ℃的硝盐浴中保温90 min，进行等温淬火处理，随后出炉空冷至室温。[14]等温淬火工艺如Fig.7-2-1所示。

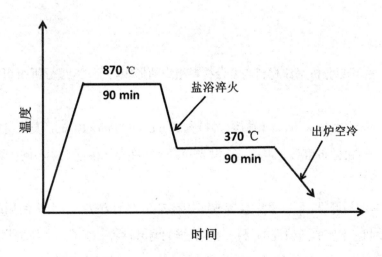

Fig. 7-2-1 The process of isothermal quenching

7.2.3.5　实验设备

本次实验用到的所有设备在Table7-2-4中列出。

Table7-2-4　The equipments used in the present study

设备名称	型号与参数
中频感应炉	60 kW, 650 V
数字维氏硬度计	HVS-30Z\LCD
光学显微镜	日本OLYMPUS-GX51
X射线衍射仪	日本理学D/MAX-2500/PC
能谱分析仪	JCM-6000
磨损试验机	日本SUGA
体视显微镜	SMZ1500
扫描电镜	JCM-6000

7.2.4　复合材料的制备

本实验利用消失模铸造和镶铸技术相结合来制备柱状碳化钨增强球铁基复合材料,并对复合材料进行等温淬火处理,最终获得奥贝球铁基复合材料,预获得复合材料增强体与基体之间达到良好的冶金结合,具有结合强度高,既硬又韧,增强体不易脱落的效果。

本实验所设计的柱状碳化钨增强表面复合材料,其碳化钨棒在复合层中均匀有序分布,这有利于减小球铁凝固时碳化钨棒与球铁热力学性能不匹配造成的影响。另外,相邻碳化钨棒之间距离太近会影响金属流动性,且增强体会引起冷铁效应,导致硬质合金与母材结合不良;距离太远又起不到有效的保护作用,因此本实验复合材料复合层中相邻碳化钨棒之间距离选定为5~6 mm(其中增强体的体积分数为3.39%,其工作表面增强体面积分数为6.78%),使复合层与基体之间的结合更加牢固,保证材料优异的抗磨性。

7.2.5　实验方法

7.2.5.1　金相组织观察

通过金相组织观察,可以分析基体材料与复合材料组织的形貌特征,这是分析组织与性能之间联系的重要环节。

将试样加工成10 mm×10 mm×10 mm(基体材料)与20 mm×20 mm×10 mm(复合材料)的尺寸,镶件→使用不同目数的砂纸粗磨→粗抛布粗磨→再精抛布抛光→腐蚀后在显微镜下观察不同倍数的金相组织。

7.2.5.2　磨粒磨损实验

为了评价某一材料的耐磨性,可以采用直接或间接模拟工况的方法。本实验使用的磨损试验机为日本SUGA型磨粒磨损试验机。本实验所研究的材料两体磨损的耐磨性评估方法:磨粒磨损实验。磨损实验测试载荷:19.6N。磨损测试试样尺寸:基体材料为50 mm×50 mm×4 mm,复合材料为50 mm×50 mm×20 mm。

7.2.5.3　实验表征方法

（1）硬度测试

硬度值是衡量材料耐磨性的重要指标之一，材料硬度按国标GB/T 231.1-2009《金属材料布氏硬度试验方法》进行测量。表面硬度测量在布氏硬度计上进行，选取直径为5 mm的淬火钢球压头，在7355N的试验力下保持12 s后进行测量。每个试样取5个点，再取平均值，即为被测试样的硬度值（HBS）。

为确定复合材料试样组织中基体、硬质合金及过渡层的硬度，在HVS-30Z型数字显微硬度计上进行显微硬度的测定。将照完金相的复合材料试样在300N的载荷下加载15 s后进行测量。从基体到硬质合金分别取3条硬度测试线，每条线取12个点作为测试点，进行测量。

（2）X射线衍射分析和SEM/EDS分析

可利用XRD分析法对材料进行物相分析，利用分析结果初步判断材料中的相组成。利用SEM分析复合材料界面结合情况，并用EDS分析其元素的分布及扩散情况。

7.3　柱状碳化钨表面增强耐磨复合材料的金相组织

复合材料是用不同的复合技术使两种或两种以上不同性能的金属通过不同的工艺方法结合在一起，使各金属发挥各自的优点，获得更优异的材料。本文主要通过简便的真空消失模工艺来制备碳化钨表面强化铁基复合材料，获得抗磨性较高的耐磨部件，以满足服役要求。

本章将对新型柱状碳化钨表面增强复合材料进行金相组织观察，并分析碳化钨与母材的结合情况，最终得出结论。

7.3.1　复合材料的组织观察及形貌分析

7.3.1.1　基体的组织形貌

将基体材料球铁与奥贝球铁进行制样、研磨、抛光，分别观察其金相组织，其试样用4%的硝酸酒精溶液进行腐蚀。

球铁在不同倍数下的组织形貌如Fig.7-3-1a, b所示。从Fig.7-3-1a中可以看出，石墨大小均匀、分布均匀，是典型的"牛眼"状球铁金相组织。Fig.7-3-1b表示放大图片，其中黑色球状区域为石墨，石墨周围白色组织为铁素体，其余呈灰黑色的区域是珠光体，结合球铁的X射线衍射分析（Fig.7-3-3b）可知，球铁基体组织主要是珠光体+铁素体。利用Axio Imaging软件对放大倍数100倍的球铁试样取5个视场进行球化率、石墨大小及物相比例的测定，最后取平均值。金相检验结果为本实验制备的球铁球化率为80%（3级），石墨评级直径为54.92 μm（6级），石墨含量约为9.47%，珠光体所占面积比例为72%左右。

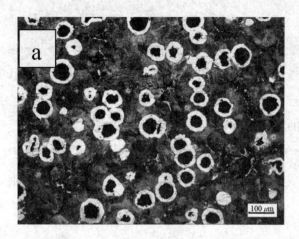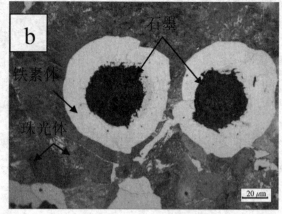

Fig. 7–3–1　Metallographic structures of SGI

　　奥氏体析出针状铁素体是在一个时间范围内完成的，在这个范围内铁素体析出量随时间延长而增加，超过这个范围将有碳化物（渗碳体）析出。奥氏体的稳定性也是在一个时间范围内增加，达到峰值后趋于下降。奥氏体含碳量高，具有较高强度，而且能提高组织的稳定性，即使铸件冷却到–120 ℃，高碳奥氏体也不会分解。[15]针状铁素体从处于过冷状态的奥氏体中析出，又受到硅的固溶强化作用，强度也得到显著提高。Fig.7–3–2为奥贝球铁不同倍数下的金相组织，该试样经过870 ℃奥氏体化，保温时间为90 min，随之在370 ℃硝盐浴中保温90 min，再空冷至室温。图中条形组织为针状铁素体，针状铁素体之间的浅色块状区域为高碳奥氏体（残余奥氏体），并结合奥贝球铁的XRD分析图谱（Fig.7–3–3a）可确定，奥贝球铁主要由针状铁素体+高碳奥氏体组成。

108

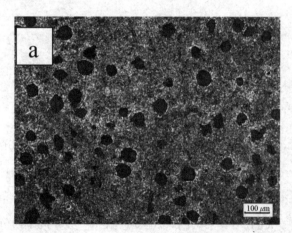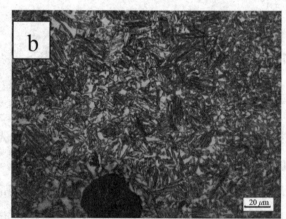

Fig. 7–3–2　Metallographic structures of ADI

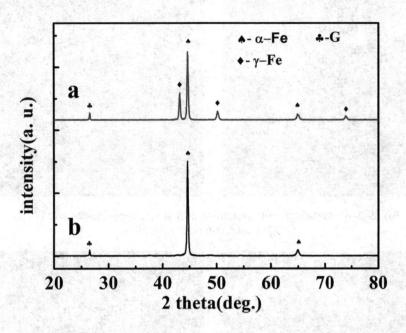

Fig. 7–3–3　X–ray diffraction spectrum of different materials
（a: ADI，b: SGI）

7.3.1.2　复合材料的组织形貌

  球铁基复合材料的组织如Fig.7-3-4a所示，在球铁金属液的热量作用下，当金属液与碳化钨棒接触时，碳化钨表层被熔融金属液热量所熔化。由于本实验球铁液浇注温度只略高于碳化钨出现液相的温度[16]，碳化钨把部分温度传给整个碳化钨棒使整个碳化钨棒的温度升高的同时，还把部分温度通过其表面散失掉，因此碳化钨棒不会大面积被融化，而是碳化钨棒表层发生局部重熔，靠近过渡层的碳化钨棒中低熔点的黏结剂钴被溶解，碳化钨颗粒扩散到周围的球铁基体中形成过渡层，过渡层中的石墨多为球状，还有部分石墨可能受碳化钨（WC）颗粒的影响，分解成颗粒状。过渡层组织较均匀，无气孔、裂纹等铸造缺陷。未受热影响的碳化钨仍然保持原有的组织形貌，如Fig.7-3-4b所示。碳化钨（WC）颗粒多为四边形或梯形，未受热影响的碳化钨原始组织区域与结合过渡层中间又形成了碳化钨组织松散的区域。Fig.7-3-5为奥贝球铁基复合材料的界面组织及未受热影响的碳化钨显微组织，奥贝球铁基体与碳化钨的结合情况等与球铁基复合材料基本相同，只是过渡层的扩散更大，过渡层的石墨多为颗粒状。经等温淬火处理后未受热影响的碳化钨显微组织也基本保持了原有的组织形貌。

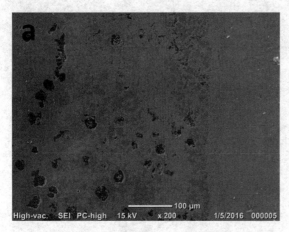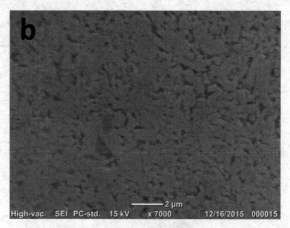

Fig. 7-3-4　Metallographic structures of SGI matrix composite and WC
（a: SGI based composite，b: WC）

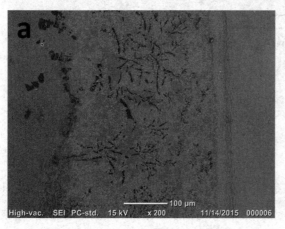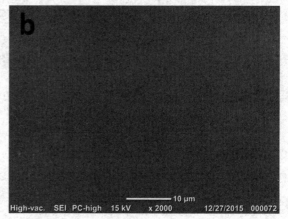

Fig. 7-3-5　Metallographic structures of ADI matrix composite and WC
（a: SGI based composite，b: WC）

7.3.2　复合材料的界面微观形貌与能谱分析

7.3.2.1　球铁基复合材料的界面微观形貌与能谱分析

球铁基复合材料的结合界面微观形貌如Fig.7-3-6所示，基体与增强体碳化钨之间出现了明显的结合过渡层，过渡层由三个区域构成（Ⅰ，Ⅱ，Ⅲ），总的宽度约为360 μm。Ⅰ区为球墨铸铁与过渡层的交界处，Ⅰ区放大图中，稀少的碳化钨颗粒分布在基体及未球化的石墨上，分布有碳化钨颗粒的基体组织较正常基体组织形貌有所改变，交界处的石墨球化不良可能是由于高温球铁液被碳化钨硬质合金激冷，硬质合金引起冷铁效应，缩短了球铁的球化及凝固时间，导致石墨球化受影响。Ⅱ区为过渡层，Fig.7-3-6Ⅱ为Ⅱ区的放大图，可看出碳化钨（WC）颗粒在基体上分布相对均匀，碳化钨（WC）颗粒形状多为四边形或梯形。石墨形状多为球状和颗粒状。Ⅲ区为碳化钨与过渡层的交界处，Fig.7-3-6Ⅲ为Ⅲ区的放大图，此区域碳化钨（WC）颗粒与正常碳化钨组织相比有明显的不同，碳化钨（WC）颗粒分布较松散。在碳化钨区域，在黏结剂钴的作用下，组织为致密的原始组织。从复合材料的界面组织及形貌基本可以判断，球铁与增强体碳化钨之间发生了良好的冶金反应并形成了明显的过渡层。

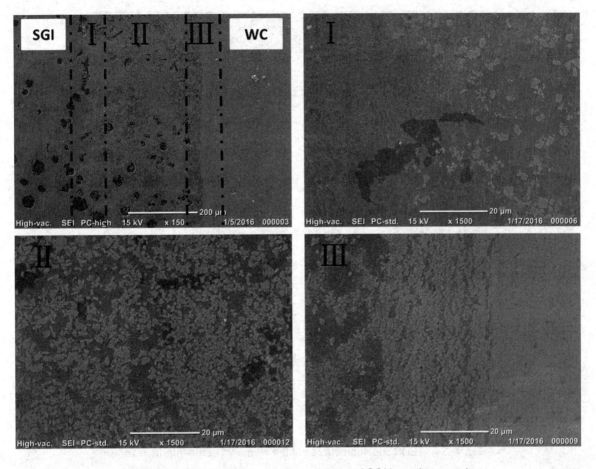

Fig. 7-3-6　The interface micromorphology of SGI based composite

　　为了进一步分析复合材料界面结合情况,对结合界面区域进行了线扫,Fig.7-3-7为球体基复合材料界面的元素能谱线扫描相应位置及结果,从扫描结果可以看出,在界面上碳化钨与基体发生了明显的元素互扩散,其中铁和钨元素的扩散现象很明显,钴元素能谱线扫描曲线趋于平滑,是由于钴元素含量少,溶解度较高,受球铁液的热作用熔解后较快扩散到基体中,碳元素能谱线扫描曲线也是较平滑,出现几个强峰是因为线扫时经过了几个石墨相。

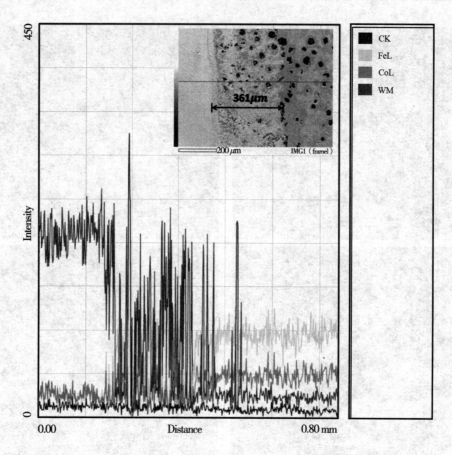

Fig. 7-3-7　EDS line scan curves of the elements of the interface zone

　　线扫描只能说明一条线上的元素扩散情况,因此为了更进一步的分析,对球铁基复合材料结合界面区域进行了元素的能谱面扫,对主要元素碳、铁、钨、钴的扫描结果如Fig.7-3-8所示,碳元素主要集中在球状及小块状区域,也就是主要以石墨形态存在,Fe主要分布在球铁基体、过渡层及增强体组织松散的Ⅲ区,从Fe元素的分布可知,Fe元素扩散良好。W主要集中在增强体及过渡层中,而Co元素在基体及增强体上分布较均匀,而量非常少。结合前面分析的复合材料组织形貌(Fig.7-3-6)可知,复合材料主要元素都有一定程度的扩散,基体与增强体之间确实形成了结合过渡层,而过渡层的存在对所制备的复合材料的组织及性能有重要的影响。

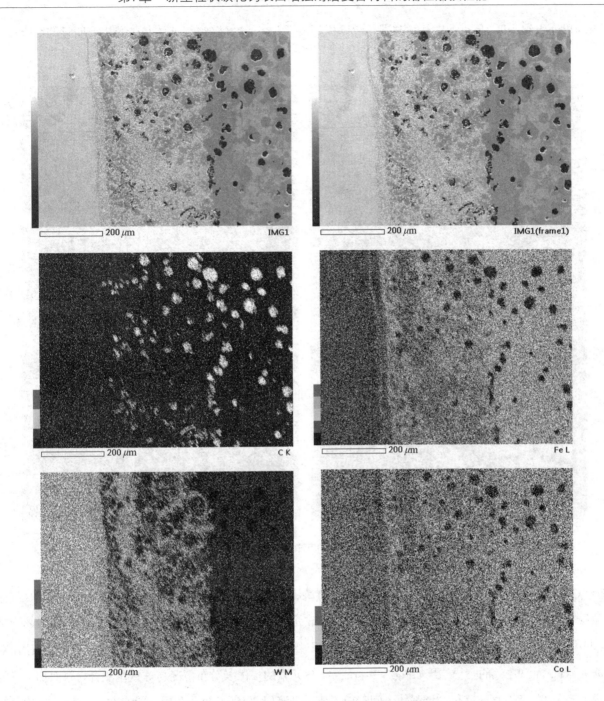

Fig. 7-3-8　EDS elements distribution of SGI based composite

7.3.2.2　奥贝球铁基复合材料的界面微观形貌与能谱分析

奥贝球铁基复合材料的结合界面微观形貌如Fig.7-3-9所示。奥贝球铁基复合材料过渡层的界面结合情况基本与球铁基复合材料的相同，同样（与球铁基复合材料），基体与碳化钨硬质合金之间出现了明显的结合过渡层，过渡层也是由（Ⅰ，Ⅱ，Ⅲ）三个区域构成，其总的宽度约为370 μm 。Fig.7-3-9Ⅰ为奥贝球墨铸铁基体与过渡层的交界处Ⅰ区的放大图，Fig.7-3-9Ⅱ为过渡层Ⅱ区放大图，Fig.7-3-9Ⅲ为增强体碳化钨与过渡层的交界处Ⅲ区放大图。等温淬火后的球铁基复合材料与铸态下的球铁基复合材料，其组织形貌及元素扩散情况基本相同。奥贝球铁基复合材料过渡层的石墨多为颗粒状，Ⅰ区及靠近Ⅰ区的基体也存在石墨球化不良现象。复合材料的界面组织表明, 奥贝球铁基体与碳化钨同样形成了良好的冶金结合, 基体与碳化钨有明显的结合过渡层。

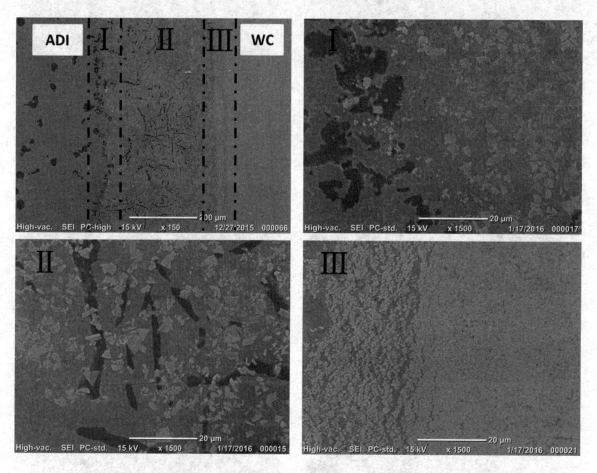

Fig. 7-3-9　The interface micromorphology of ADI based composite

　　奥贝球铁基复合材料界面区域的元素能谱线扫描结果及相应位置如Fig.7-3-10所示。从结果可知, 在界面区域碳化钨与基体同样发生了元素互扩散, 其中Fe和W元素扩散现象非常明显, 而Co, C元素线扫曲线趋于平滑, 其中C元素出现几个强峰也是因为线扫时经过了几个石墨相。又对奥贝球铁基复合材料进行了面扫描, 对主要元素C, Fe, W, Co的扫描结果如Fig.7-3-11所示。各元素的扩散及分布情况与球铁基复合材料面扫得到的结果大致相同, 复合材料基体与增强体之间形成了结合过渡层, 保证了两个金属材料(母材与增强体)冶金结合并获得更优异的性能。

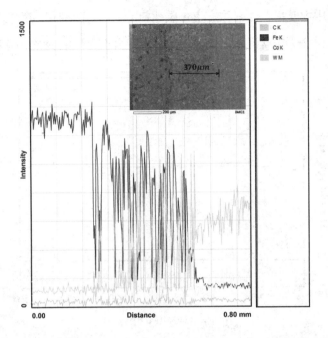

Fig. 7-3-10　EDS line scan curves of the elements of the interface zone

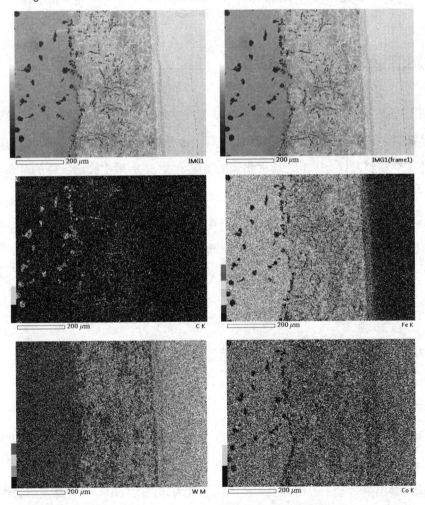

Fig. 7-3-11　EDS elements distribution of ADI based composite

通过对两个复合材料结合部位的XRD分析（Fig.7-3-12）可知，奥贝球铁基复合材料（Fig.7-3-12a）及球铁

115

基复合材料（Fig.7-3-12b）中主要存在铁素体α-Fe、碳化钨（WC）这两种相。除此之外，奥贝球铁基复合材料中还含有γ-Fe，两个复合材料还应该有石墨相和渗碳体，可能由于含量太少，XRD没能检测出来。

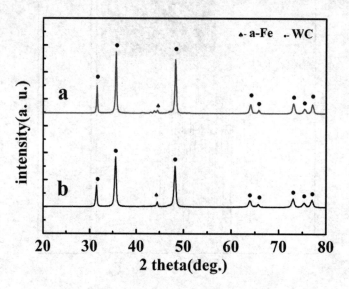

Fig. 7-3-12 X-ray diffraction spectrum of different composite materials
（a: ADI based composite, b: SGI based composite）

7.4 复合材料的磨粒磨损性能研究

磨粒磨损是由于硬质物料或突出物与表面相互摩擦使材料发生损耗的现象。电力、矿山、建筑、煤炭、机械等领域普遍存在工况复杂、磨损严重的现象，尤其是由于磨粒磨损，每年耗损大量的钢铁耐磨件。硬质合金增强铁基耐磨复合材料恰好可以适合这种恶劣的工况要求，可有效提高耐磨件的可靠性及使用寿命。因此，研究耐磨复合材料的磨粒磨损特性有极大的实际意义。

由磨损所引起的材料损失称作磨损量，磨损的大小可用磨损量来表示，它的倒数即为耐磨性。本实验采用测量磨损量（被磨去的质量m）的方法对材料进行耐磨性测试。

本章将对碳化钨表面增强奥贝球铁基复合材料进行磨粒磨损试验，并在相同条件下，分别与基体球铁、奥贝球铁及球铁基复合材料的磨粒磨损性能进行比较，然后分析结果，得出结论。

7.4.1 磨粒磨损实验结果

7.4.1.1 基体磨粒磨损性能对比

球铁与奥贝球铁经过SUGA型磨料磨损试验机的磨损后，磨损往复次数与磨损量之间的关系如Fig.7-4-1所示。在相同的磨损往复次数下，对两种基体材料磨损量进行比较，发现2 000次往复磨损后，球铁的磨损量为0.27 g，奥贝球铁的磨损量为0.18 g，即奥贝球铁的耐磨性为球铁的1.5倍。奥贝球铁耐磨性提高的一个重要原因可能是奥贝球铁微观组织中确定的残余奥氏体经历了相变硬化，奥氏体转变成为马氏体，大大增加了表面硬度，从而总会有一层高硬度的抗磨工作表面。

116

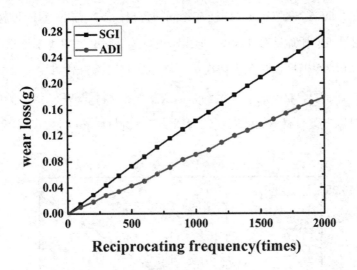

Fig. 7-4-1　Abrasion results of different materials

7.4.1.2　基体与复合材料磨粒磨损性能对比

为了进一步提高球铁的磨粒磨损性能，制备了碳化钨表面增强球铁基复合材料，并对制备好的试样进行磨粒磨损实验。在相同的磨损往复次数下，对球铁及碳化钨增强后的球铁基复合材料磨损量进行比较，磨损往复次数与磨损量之间的关系如Fig.7-4-2所示。发现，球铁磨损量为球铁基复合材料磨损量（0.13 g）的2倍，也就是说，球铁基复合材料的耐磨性是球铁的2倍，说明在磨粒磨损过程中，复合材料组织中硬质合金增强区域突出于球铁基体，且由于界面结合良好，其碳化钨增强体保护基体和球铁基体支撑增强体的协同效应，是球铁基复合材料耐磨性优越的内在原因。

相比球铁，耐磨性更好的奥贝球铁的磨损量是球铁基复合材料的1.4倍。因此，可通过对奥贝球铁进行碳化钨表面强化处理，制备碳化钨表面增强复合材料，来提高它的耐磨性。

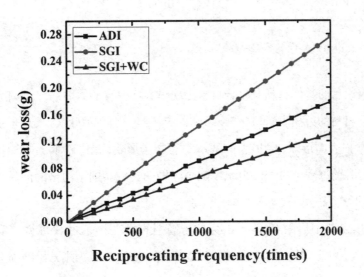

Fig. 7-4-2　Abrasion results of different materials

7.4.1.3　不同基体及复合材料磨粒磨损性能对比

球铁、奥贝球铁、球铁基复合材料及奥贝球铁基复合材料经过SUGA型磨料磨损试验机的磨损后，磨损往

117

复次数与磨损量之间的关系如Fig.7-4-3所示。同样在相同的磨损往复次数下，对四种材料磨损量进行比较，发现球铁基复合材料磨损量虽然比球铁减少50%，但由于奥贝球铁耐磨性更优于球铁，它的磨损量是奥贝球铁基复合材料磨损量的1.4倍。磨损量从多到少依次为球铁、奥贝球铁、球铁基复合材料及奥贝球铁基复合材料。通过最终计算，奥贝球铁基复合材料的磨损量约为球铁基复合材料的69%、基体奥贝球铁的50%、基体球铁的33%，即设计的奥贝球铁基复合材料的耐磨性约为球铁基复合材料的1.4倍、基体奥贝球铁的2倍、基体球铁的3倍。

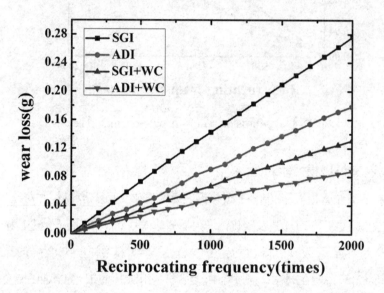

Fig. 7-4-3　Abrasion results of different materials

7.5　实验结果分析

7.5.1　磨损过程分析

7.5.1.1　磨损表面观察

为了进一步分析淬火前后的基体材料及复合材料出现耐磨性差异的原因，本实验使用扫描电镜观察了淬火前后的基体材料及复合材料的磨损表面，Fig.7-5-1到Fig.7-5-4分别表示球铁、奥贝球铁、球铁基复合材料及奥贝球铁基复合材料的磨损表面形貌，磨损表面基本都有犁沟形貌，是由砂纸上的SiC磨粒在磨损过程中划损表面所造成的，磨粒在滑动过程中将表面组织犁削至沟槽两边或切削至沟槽的前端，使组织发生变形、断裂，最终以切屑或犁屑的形式脱离表面。

球铁的磨损过程如Fig.7-5-1所示，Fig.7-5-1a为还未磨损的球铁表面。由于球铁硬度较低，主要为塑性变形（材料在摩擦力的作用下，发生塑性变形）和小的黏着撕裂。Fig.7-5-1b为刚开始磨损的试样表面，已出现明显的犁沟。在磨损过程中（如Fig.7-5-1c所示），在犁沟的沟槽边缘存在大量卷曲、变形成未脱落的磨屑及较大的裂纹。球铁组织中石墨非常软，所以磨损时容易变形及脱落。石墨周围是铁素体，铁素体也是软韧相，硬度低，遇到硬度高的碳化硅磨粒时，被切削至石墨上（Fig.7-5-1d）。由于组织塑性较好，部分被犁削致沟槽两侧的金属组织仍然附着在表面, 与基体保持良好的连接, 但在反复切削作用下, 组织仍会发生断裂、脱落。球铁的磨损方式主要为犁沟、黏着、塑性变形。

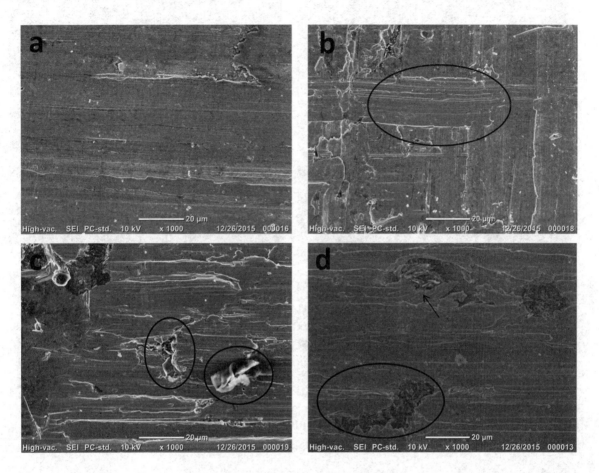

Fig. 7–5–1　Worn surface morphology of SGI

　　Fig.7–5–2a为奥贝球铁的未磨损表面, 相较于球铁未磨损表面, 其表面平滑许多。Fig.7–5–2b为磨损初期的表面样貌, 材料开始被剥离, 材料被反复切削后开始剥落。在Fig.7–5–2c中, 在滑动过程中组织被磨粒犁削至沟槽两边, 使组织发生变形并开始断裂, 最终以切屑或犁屑的形式脱离表面。如Fig.7–5–2d所示, 奥贝球铁组织中的石墨虽然软, 但是奥贝球铁基体组织主要由针状铁素体+高碳奥氏体组成, 石墨周围不是单相的铁素体, 而残余奥氏体在磨粒磨损过程中发生相变强化, 转变为硬相马氏体, 提高了组织硬度, 减少了塑性变形, 因此石墨变形、脱落, 金属流入等现象明显得到改善。

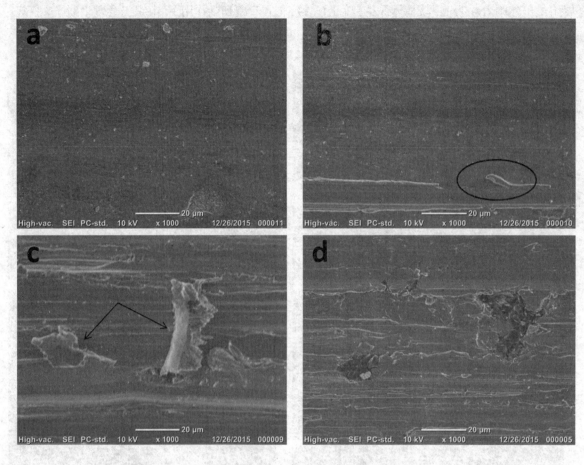

Fig. 7–5–2　Worn surface morphology of ADI

Fig.7–5–3为球铁基复合材料的磨损表面形貌。Fig.7–5–3a为两个相邻硬质合金中间基体球铁的磨损形貌,由于磨损时硬质合金对基体的保护作用,石墨形状较球铁基体明显平整。Fig.7–5–3b为磨损(摩擦面入口处)初期的基体磨损表面,磨损面较为平整,基体黏着磨屑较少。Fig.7–5–3c为复合材料的界面磨损形貌,可知基体与硬质合金的结合处磨损时产生了较多的裂纹,这有可能是复合材料结合处结合不良及基体断裂韧性降低所致。

Fig. 7–5–3　Worn surface morphology of SGI based composite

Fig.7–5–4为奥贝球铁基复合材料的磨损表面形貌。Fig.7–5–4a是两个相邻硬质合金中间基体奥贝球铁的磨损形貌,犁沟及塑性变形明显少于球铁基复合材料,这是由于等温淬火处理后球铁硬度有所提高所至。Fig.7–5–4b为磨损面入口的基体磨损形貌,可清晰地看到被犁削至沟槽两侧断裂的犁屑及切削至沟槽前端翘起的切屑,

与球铁基复合材料相比犁沟显著减少。Fig.7-5-4c为奥贝球体基与硬质合金结合界面磨损形貌,由于硬质合金在磨损过程中对基体的保护作用及基体对增强体的支撑作用,交界处较球铁基复合材料明显平滑且基本没有裂纹,结合良好。

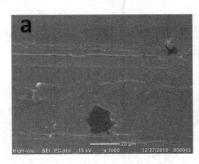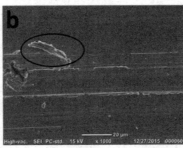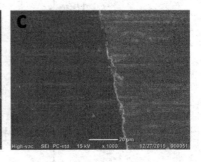

Fig. 7-5-4　Worn surface morphology of ADI based composite

经过与砂纸一定次数的往复磨损后,四个试样表面均形成程度不同的犁沟。复合材料磨损时划痕在经过硬质合金时被阻断或减弱,这种阻断既缩短了犁沟的长度,又减少了变形次数,降低了对材料的磨损,使材料表现出良好的耐磨性。

7.5.1.2　硬度测试结果

不同基体与不同复合材料在磨损过程中表现出了不同的抗磨损能力。为了进一步分析磨损性能,本实验分别对球铁、奥贝球铁进行了布氏硬度测试,对球铁基复合材料及奥贝球铁基复合材料进行了维氏硬度测试。实验结果如Table7-5-1, Fig.7-5-5及Fig.7-5-6所示。从Table7-5-1可以看出,球铁的硬度为260.4 HBS。奥贝球铁的硬度为315.9 HBS,比球铁的硬度高21%,奥贝球铁中金属基体的实际硬度要比测量的硬度高,因为通常硬度测量方法测量的是基体和石墨的平均硬度。原始硬质合金的显微硬度为1 286.5 HV。

在球铁基复合材料上,为了提高准确度,从基体到硬质合金分别取三条硬度测试线,每条线取12个点作为测试点进行测量,得到从基体过渡到硬质合金的硬度曲线,如Fig.7-5-5所示。对奥贝球铁基复合材料同样取三条硬度测试线进行硬度测量,结果如Fig.7-5-6所示。可以看出由于复合材料过渡层的存在,球铁及奥贝球铁基复合材料硬度曲线过渡都非常明显。碳化钨棒的中心区域基本保持其原始硬度(1 286 HV),说明等温淬火处理只会改变基体的组织及性能,对硬质合金的组织及硬度基本没有影响。过渡层的硬度较低,复合材料中球铁基体的硬度比较单一。球铁材料的硬度有所提高,这是由于碳化钨(WC)的扩散,基体的含碳量升高,致硬度提高。两个硬度曲线能更好地说明球铁及奥贝球铁基复合材料的基体与增强体结合良好。

Table7-5-1　The hardness of different materials

材料	硬度值(HBS)					平均值
	1	2	3	4	5	
球铁	256	256	263	259	268	260.4
奥贝球铁	309.5	312.5	327.5	316.4	313.7	315.9
硬质合金	硬度值(HV)					平均值
	1 298.9	1 260.6	1 286.9	1 275.1	1 311.2	1 286.5

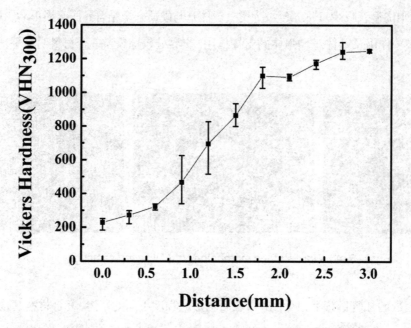

Fig. 7-5-5   The curve of microhardness for the transition layer of SGI matrix composite

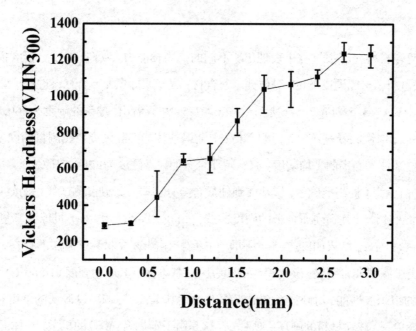

Fig. 7-5-6   The curve of microhardness for the transition layer of ADI matrix composite

7.5.1.3   磨损前后的 XRD 分析

对球铁（Fig.7-5-7）与奥贝球铁（Fig.7-5-8）进行磨损前后的XRD分析, 球铁与奥贝球铁都是石墨先变形或脱落。奥贝球铁（Fig.7-5-8）磨损后组织中的奥氏体明显减少, 这是由于奥贝球铁组织中存在的残余奥氏体在磨损过程中转变为形变马氏体, 发生相变强化, 提高了硬度, 并增强了耐磨性能。

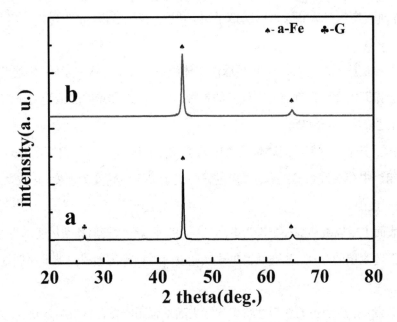

Fig. 7–5–7　X–ray diffraction analysis on SGI before and after wear
（a：before test，b：after test）

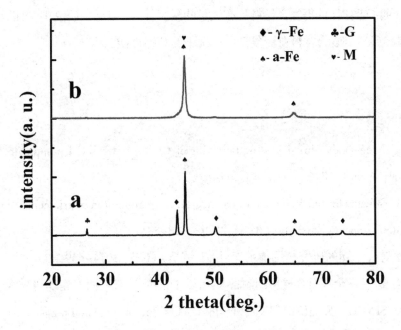

Fig. 7–5–8　X–ray diffraction analysis on ADI before and after wear
（a：before test，b：after test）

7.6　结论

从降低成本、提高工作性能和使用寿命考虑，本课题在已有基础上，结合真空消失模铸造工艺制备出以WC-Co硬质合金棒为增强体、奥贝球铁为基体的柱状碳化钨表面增强铁基复合材料。复合材料较基体在磨粒磨损条件下表现出了更好的耐磨性能。使用柱状碳化钨表面增强耐磨复合材料制造的破碎机锤头及衬板使用寿命会更长，成本也会相应降低很多。

本文主要研究内容是对比研究球铁、奥贝球铁、球铁基复合材料及奥贝球铁基复合材料。通过对基体的金相组织、复合材料界面的物相组成和元素的扩散情况、硬度、磨粒磨损性能的测定，以及磨损表面的分析等，得出以下结论：

1. 球铁基体组织主要是珠光体+铁素体。奥贝球铁基体组织主要由针状铁素体+高碳奥氏体组成。

2. 在复合材料结合界面主要元素都有一定程度的扩散，且元素Fe和W扩散最为明显，基体与增强体碳化钨之间确实形成了结合过渡层，结合牢固。

3. 所设计的奥贝球铁基复合材料的耐磨性约为球铁基复合材料的1.4倍、基体奥贝球铁的2倍、基体球铁的3倍。由于奥贝球铁组织中存在残余奥氏体，磨损过程中残余奥氏体转变为形变马氏体，发生相变强化，提高了硬度，并增强了耐磨性能。

4. 被测试样磨损表面主要以微切削和犁沟为主，其磨损机理以显微切屑和犁沟变形为主。

5. 复合材料中由于过渡层的存在，球铁及奥贝球铁基复合材料硬度曲线过渡非常明显，保证了复合材料具有优异的耐磨性能。

通过本课题的研究发现，所制备的新型柱状碳化钨表面增强耐磨复合材料中高硬度、高耐磨性的碳化钨凸出于奥贝球铁基体，对基体的磨损起着有效的保护作用；同时，奥贝球铁基体又有效支撑着增强体碳化钨，在二者相互协同作用下抵抗磨损，使得复合材料具有优异的耐磨性。在以后的研究中，可以对奥贝球铁基复合材料进行冲蚀磨损实验，研究其复合材料在更复杂的工况下的磨损性能。

参考文献

[1] LI Y, GAO Y. Three-body abrasive wear behavior of CC/high-Cr WC I composite and its interfacial characteristics [J]. Wear, 268（3-4），2010: 511-8.

[2] PRAMANIK A. Effects of reinforcement on wear resistance of aluminum matrix composites [J]. Transactions of Nonferrous Metals Society of China, 26（2），2016: 348-58.

[3] 张云, 龚文邦, 刘欢. ADI的研究应用前景探讨 [J]. 铸造,（05），2014: 439-43.

[4] 王双成. 硬质合金/钢双金属复合材料的制备及组织、性能研究 [D]. 西安建筑科技大学, 2008.

[5] KATIYAR P K, SINGH P K, SINGH R, et al. Modes of failure of cemented tungsten carbide tool bits（WC/Co）: A study of wear parts [J]. International Journal of Refractory Metals and Hard Materials, 54.2016: 27-38.

[6] 蔡美, 王双成, 许云华, et al. 硬质合金/钢双金属复合材料的组织与性能研究 [J]. 硬质合金,（04），2008: 203-7.

[7] 鲍崇高, 侯树增. 硬质合金局域化增强铁基表层复合材料的制备与磨损性能研究 [M]. 第十四届全国耐磨材料大会论文集. 洛阳: 中国金属学会特殊钢分会耐磨材料学术委员会. 2015.

[8] 刘亚民, 陈振华, 魏世忠, et al. 硬质合金-球墨铸铁复合铸造 [J]. 河南科技大学学报（自然科学版），（01），2004: 23-5.

[9] 李固成. 高铬合金铸铁/硬质合金耐磨复合材料溜槽衬板的开发与试验研究 [M]. 2014中国铸造活动周.

郑州；中国机械工程学会. 2014.

[10] 姚圣高, 沈仁才, 王猛. 一种镶铸硬质合金耐磨衬板生产制造工艺, CN103447503A [P/OL]. 2013–12–18.

[11] 鲍崇高, 侯书增, 邢建东, et al. 一种硬质合金/高铬合金基耐磨复合材料的制备方法, CN102380605A [P/OL]. 2012–03–21.

[12] 章舟, 陆国华, 刘中华. 消失模白模制作技术问答 [M]. 北京: 化学工业出版社, 2012.

[13] 邓宏运, 王春景, 章舟. 等温淬火球墨铸铁的生产及应用实例 [M]. 北京: 化学工业出版社, 2009.

[14] 侯宁, 戴秋莲, 骆灿彬, et al. 消失模铸造制备奥–贝球铁CBN砂轮胎体 [J]. 特种铸造及有色合金, （03）, 2014: 290–3.

[15] 郝石坚, 宋绪丁. 球墨铸铁 [M]. 北京: 化学工业出版社, 2014.

[16] 王国栋. 硬质合金生产原理 [M]. 北京: 冶金工业出版社, 1988.

第8章　总结

8.1　本课题的研究成果

高炉喷煤装置管道和钟形料斗中发生的冲蚀磨损,是指在煤粉和矿石粉的撞击下管道内壁或者料斗倾斜部位受到的磨损。这种冲蚀磨损因为其构造极其复杂,在实际工作中,无论是从管道的外观观察还是从高炉上方观察料斗的磨损过程都是不可能的。在冲蚀磨损的研究方面,人们对一般构造材料基本的冲蚀磨损机理较为了解,通过分析其有效点和问题点得知,各种材料因其组织及热处理条件的不同表现出自己独特的冲蚀角度依赖性。

球状碳化物铸铁材料因其组织内含高硬度的VC,比起其他现有的钢铁材料,具有极好的抗冲蚀磨损特性。另外,这种材料不存在冲蚀角度依赖性,在任一角度都具有良好的抗冲蚀磨损特性。在其基础上利用碳化钨(WC)增强表面进一步增加了其抗冲蚀磨损性。

8.2　存在的问题及展望

本文以了解球状碳化物铸铁及表面改性材料的冲蚀磨损特性为目的,应用粉粒体的气固两相流进行冲蚀磨损试验,对磨损表面进行宏观和微观观察,同时观察磨损表面附近的断面,在此基础上上评价材料的冲蚀磨损特性。球状碳化物铸铁材料表现出了良好的耐冲蚀磨损特性。这应该与组织中析出的碳化物的球状化有关。因为球状碳化物的存在,应力集中得到分散,能量被吸收,耐冲蚀磨损特性增加。另外,从角度依赖性消失的理由考虑,球状碳化物材料既能抵御低角度的切削磨损,又能承受高角度的变形磨损。然而,目前因为容易出现铸造缺陷、碳化物的偏析等,在材料组织内均匀析出碳化物有些困难。期待碳化物的均匀分散技术及材料缺陷改善技术的发展。

球状碳化物铸铁的磨损试验,本研究只在常温下利用粉粒体的固气两相流进行了冲蚀磨损试验。但在实际情况下,磨损条件是多种多样的,有必要在流体粉粒体体系中及高温条件下进行冲蚀磨损试验。

本研究采用的冲蚀粒子流速为100 m/s,单位冲蚀粒子量为不定型石英砂4~5 g/s,钢砂37 g/s。石英砂的硬度为1030 HV,钢砂的硬度为450 HV。这些条件是否适用于各种实际情况,有待于讨论。

球状碳化物的冲蚀磨损没有冲蚀角度依赖性。为了解其机理,本研究应用了材料表面的宏观和微观观察、断面观察及断面的连续观察等手段。得出的结论是,因为球状碳化物的存在,应力集中得到分散,能量被吸收。同时,因为球状碳化物的等方向性,冲蚀粒子由低角度向高角度变化过程中所产生的能量是不变的。

另外,材料基体因容易发生加工硬化,材料基体与VC的结合性增加,VC的脱落受到抑制,冲蚀依赖性逐渐消失。虽然本文建立了冲蚀磨损的基本模型,但还未进行冲蚀磨损表象的力学解析。采用3次元解析软件以基本模型为基础,用有限要素法进行模拟是有必要的。考虑在下一阶段的工作中完成这项研究。

本研究阐明了球状碳化物在固气流中的冲蚀磨损机理。球状碳化物铸铁具有良好的耐冲蚀磨损特性,目前已得到实际应用,并在很多场合得到好评。然而冲蚀磨损涉及范围极其广泛,所以关于球状碳化物铸铁冲蚀磨损方面遗留下的研究课题有必要继续研究。

Available online at www.sciencedirect.com

ScienceDirect

Wear 264 (2008) 947–957

www.elsevier.com/locate/wear

Erosive wear characteristics of spheroidal carbides cast iron

Xinba Yaer[a], Kazumichi Shimizu[a,*], Hideto Matsumoto[b],
Tadashi Kitsudo[c], Tadashi Momono[a]

[a] Department of Material Science and Engineering, Muroran Institute Technology, Muroran 050–8585, Japan
[b] Sankyo. Co. Ltd., Osaka 555-0001, Japan
[c] Technology Research Institute of Osaka Prefecture, Izumi, Osaka 594-1157, Japan

Received 11 April 2006; received in revised form 7 May 2007; accepted 3 July 2007
Available online 31 August 2007

Abstract

In the present investigation, the authors reported erosive wear property of spheroidal carbides cast irons (SCI). SCI was obtained at the temperature of 1700–1800 °C by adding about 10% vanadium to crystallize vanadium spheroidal carbides in the structure. Erosive wear tests were performed on high manganese cast iron with spheroidal carbides (SCI–VMn), high V–Cr–Ni cast iron with spheroidal carbides (SCI–VCrNi) and white cast iron (WCI) by using a shot-blasting machine. Erosion damage was evaluated by the removed material volume at impact angle range from 10° to 90°. The surface metal flow in vertical section was also observed. Surface morphology of each material was characterized with scanning electron microscopy (SEM). The effect of impact angle, and differences in wear features of specimens were discussed as mechanism of erosive wear. Experiment showed that the erosion rate for SCI is about 1/4 to 1/5 of that for WCI. The hardness of test surface of SCI–VMn increased from the initial 504 to 804 HV after 3600 s of blasting. It showed that austenite in the surface structure has been transformed to martensite, which hardened the surface and lowered the erosion rate. Similarly, work-hardening effect also occurred on SCI–VCrNi to make its surface hardness increased to 478 HV. It was shown that SCI series have excellent erosive wear resistance and they are expected to find wide application as wear-resistant materials.

Keywords: Erosive wear; Spheroidal carbides; Erosion rate; Work hardening

1. Introduction

Erosive wear, or the sand erosion phenomenon, has been a serious problem in many industrial systems since last century. Recently, big accidents have occurred due to the erosive wear phenomena. Take the pipe damage in the Mihama power station for example; the wall of pipe in the factory was thinned down by erosive–corrosive wear, resulting in the accident. Similarly, the erosion phenomena can cause serious problems; mainly at bended sections of pipe, valve, turbine blade, the fan in a pneumatic conveying system, and even the blade of helicopter, etc. [1–5] may be damaged.

Instances of sand erosion occur in secondary refining and smelting reduction equipment at iron and steel plants. When the dispersed particles such as dust coal, powdered mineral, etc.,

were blown into the melted pig iron through blowpipes, the erosion damage occurred at the bended section of the pipe. If these kinds of pipe systems were eroded and were perforated in places by gas entrained particles that penetrate the inner surface, they would likely result in causing serious industrial accidents.

Mechanism of erosive wear in pipe was very complex, and there occurred a large amount of damages by erosive wear. Nevertheless, it could not predict which part in the pipe was most damaged and how much life span the pipe had left. So when regular maintenance was performed, the inner wall of the pipe to be thinned down was built up through welding in order to avoid causing accidents. But this kind of prevention is only temporary; damage will occur again in 2–3 weeks. Therefore, to further prevent the accident from occurring, important research is being done about the development of wear-resistant material and estimation of life span during erosion.

The dominant factors that influence erosion are mainly associated with the mechanical properties of objective materials, especially, the hardness of the materials. In addition to the

* Corresponding author. Fax: +81 143465651.
E-mail address: shimizu@mmm.muroran-it.ac.jp (K. Shimizu).

hardness, the shapes, hardness and sizes of impact particles, impact speed, and impact angle of particles also affect the amounts of materials removed and erosion characteristics of materials [9–12].

Look over previous works of other researchers, many erosive wear testing were proposed to explain the mechanism of objective materials. From the research on the effect of impact angle, it is generally concluded that erosive wear of material depends on the impact angle of particles and a material has weak angle and strong angle to particles' impacting.

The erosion of ductile material stroke by angular abrasive particles was formulated by Finnie [1] at 1960. On the basis of two dimensions cutting theory, Finnie established expressions of erosive wear amount to approximately explain the influence of impact angle and impact speed of abrasive on erosion. But the predicted value underestimates erosion at high-impact angles disagree the experimental value of erosion at 90°. Bitter [2,3] developed the Finnie's theory with very comprehensive study of the problem, and recognized that it is necessary to exceed threshold conditions in order to cause damage. The dependence on impact angle was explained by the theory of what was termed "cutting" erosion occurring at glancing impact whilst "deformation" erosion predominated at angles close to the normal. Neilson and Gilchrist [4] subsequently simplified Bitter's analysis, and then Tilly [5–8] had introduced the incubation effect in the influence of impact angle and presented a two-stage mechanism on ductile erosion. Hutchings and Winter [9,10] introduced the ploughing method to explain the erosive wear mechanism. As is known to all, these studies laid the foundation of erosion mechanism. However, it is found that impact angle dependency of erosive wear differs from one to another not only because of characteristic of target materials but also because of property of impact particles and impact condition. Accordingly, it seems unreasonable that direct application of the results of previous work in the erosive wear assessment of other industrial materials. And then recently, many of researchers [11–14] mainly aimed at the erosive wear of composite material, superalloy, hard surface coatings, high chromium white cast iron, and so on; among these, high chromium white cast irons have been defined previously to good wear-resistance materials in abrasive condition, however, the presence of dendrite or flaky shaped carbides embedded in matrix drew them occur stress concentration easily to limit their wear resistance. Therefore, we developed the cast irons with spheroidal carbides embedded in matrix, and there has little attention on the study of these kinds of cast irons.

Spheroidal carbides cast irons were obtained in the temperature of 1700–1800 °C by adding about 10% vanadium to crystallize vanadium spheroidal carbides in the structure.

The object of research described in this paper is to evaluate the erosive wear properties of two kinds of spheroidal carbides cast irons, which are expected to be effective wear-resistant materials. In this research, from this point of view, spheroidal carbides cast irons were employed, as well as a number of high chromium carbides cast irons which had been used for hydraulic runner castings and the parts of hydraulic equipments for comparison.

2. Experimental details

2.1 Testing machine

A shot-blasting machine was used to test the erosion of specimens in this research. The schematic view of the testing machine is showed in Fig. 1.

2.2 Materials

Spheroidal carbides cast irons selected for the present investigation was high manganese cast iron with spheroidal carbides (SCI–VMn) and high V–Cr–Ni cast iron with spheroidal carbides (SCI–VCrNi). Also two kinds of tempered high chromium white cast iron (mass% of Cr is 26% and 17%; 26Cr (tempering), 17Cr (tempering)) were also used for comparison. The chemical compositions of the alloys are given in Table 1.

2.3 Metallography

Materials used in this research were characterized by optical microscope. The specimens measuring 10 mm × 15 mm were used for metallographic processing. The grinding was finished with silicon carbide paper to 1000 grit. The polishing was carried out on the diamond-polishing machine to obtain a surface finish of 1 μm. Next the polished specimens of objective materials were etched by natal (the amount of HNO_3 was 5%).

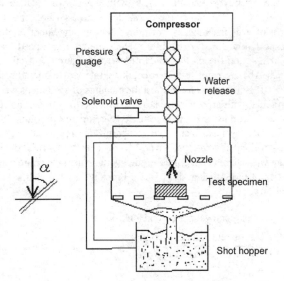

Fig. 1. The schematic view of shot-blasting machine.

Table 1
Chemical composition of various specimens

	C	Si	Mn	Cr	V	Ni	P
17Cr (tempering)	2.97	0.5	0.72	17.28	–	–	0.03
26Cr (tempering)	2.54	0.46	0.8	26.28	–	0.1	0.03
SCI–VCiNi	2.84	0.95	0.61	17.3	9.34	9.2	–
SCI–VMn	2.92	0.57	12.9	–	11.9	0.95	–

129

Table 2
Material characteristics of alloys

Material	Microstructure	Initial Vickers hardness (HV)	Micro-Vickers hardness (HV)	Carbide volume fraction
17Cr (tempering)	Eutectic M_7C_3, perlite matrix	510	Carbides 1500 Matrix 230	28–32%
26Cr (tempering)	Eutectic $M_{23}C_6$, perlite matrix	478	Carbides 1130 Matrix 230	28–32%
SCI–VCrNi	VC, eutectic M_7C_3, austenite matrix	399	VC 2300 M_7C_3 1500 Matrix	VC: 14–18%, M_7C_3: 5–8%
SCI–VMn	VC, austenite matrix	533	VC 2300 Matrix	20–24%

Finally, electronic microscope was used to evaluate the various phases in the microstructure of selected specimens. The materials characteristics of the alloys are given in Table 2 and their metallographic structures of the alloys are shown in Fig. 2.

2.4 Erosion test

A shot-blasting machine was used to test the erosive wear of target materials in the present investigation. Angular (irregularly shaped) silica sand (2 kg) with average diameter 408 μm, Vickers hardness 1030 HV were used as impact particles. The impact particles were changed after each test because the particles themselves also were eroded, changing the size of the particles. Specimens selected for the present investigation were cast iron with spheroidal carbides, high chromium cast iron. The specimens measuring 50 mm × 50 mm × 10 mm were used for erosion test. The specimens were mounted into the test stage directly below the nozzle with a vertical distance of 50 mm from the end of the nozzle to the test surface into the erosion test machine by changing their impingement angles respectively 30°, 60°, and 90°. The examined air speed was 100 m/s, and the particle feed rate was measured with about 4 g/s. All the erosion tests were conducted at room temperature in 3600 s. Before and after the test the amounts of specimens were weighed with an electronic scale to prepare for measurements of erosion rate. It is more accurate to compare by volumetric loss than by mass loss when comparing the removed materials from specimens that have different densities. The erosion rate was calculated from the eroded volume using average densities of target materials [17–19]. Erosion rate was defined as follows:

volumetric removal per second (cm^3/s)

$$= \frac{\text{mass removal per second (g/s)}}{\text{average density (g/cm}^3)}$$

erosion rate (cm^3/kg)

$$= \frac{\text{volumetric removal per second (cm}^3/s)}{\text{mass amounts of impact particles per second (kg/s)}}$$

The reason why was the volumetric removal divide by mass amounts of impact particles is that the mass removal per second should depend on the total amounts of particles that impact on target material in that period. It was considered that the more the amounts of particles impact on surface of material the more the material removal increased. So it is more reasonable to determine the value at amount of impact particles per second as erosion rate.

3. Observation of experimental result

Fig. 3(a)–(c) shows the mass loss of specimens as a function of erosion time for irregularly shaped silica sand particles at incidence angle of 30°, 60°, and 90°, respectively. The figures reveal that for three angles, after initial stage (from results of relevant works for erosion there was evidence of a short induction period before any mass loss was observed [5–8]), the mass loss is approximately linear with erosion time. It can also be observed that SCIs show far flatter lines at all of angles than WCIs. It indicates that SCIs have higher erosive wear resistance and have longer life span than WCIs.

The erosion test was performed on SCI–VMn with steel grits. In the test with spherical steel grit, SCI–VMn has not been eroded and shows no clear peak. From this point, angular silica sand that is harder than steel grits was used to carry out erosion test on the materials. Fig. 4 shows the result of erosion rate in SCI–VMn by both steel grit and silica sand. It can be clearly seen that although SCI–VMn was almost not eroded in erosion test with steel grit, there showed erosion in SCI–VMn when the hard and angular silica sand was used as impact particles. We draw conclusions that the removed mass changes dramatically if the impact particles were changed.

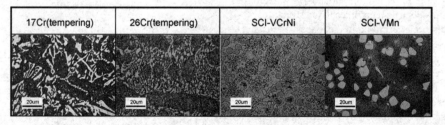

Fig. 2. Metallographic structures of the alloys.

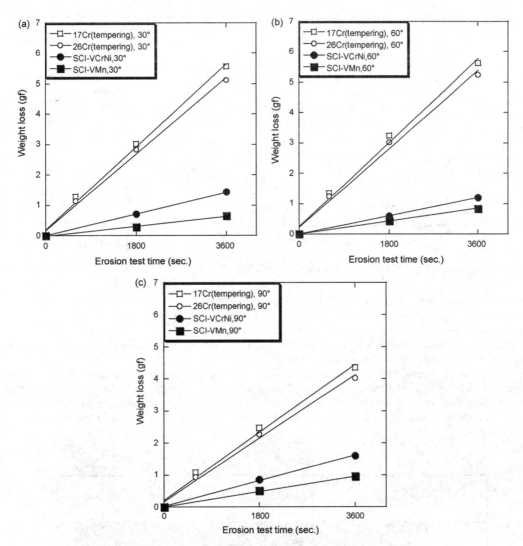

Fig. 3. (a) Erosion test time vs. mass loss in specimens at 30°; (b) erosion test time vs. mass loss in specimens at 60°; (c) erosion test time vs. mass loss in specimens at 90°.

The erosion test was also performed on other materials with silica sand. The result is shown in Fig. 5. From the figure we can see clearly that SCI–VCrNi, as well as SCI–VMn, shows best wear-resistant properties. The expected wear-resistant material of 26Cr (tempering) and 17Cr (tempering) that are very hard and have flaky carbides in structure show peak at 60°, and at 30°, 90° the erosion rates were both dramatically declined. However, the two kinds of high chromium cast iron showed a considerably higher erosion rate than the two kinds of spheroidal carbide cast irons. Erosion rate of SCI–VCrNi shows slightly higher erosion rate at 90° than 30° and 60°, but the variation is extremely small. SCI–VMn exhibits lowest erosion rate. And it is observed from graph clearly that SCI–VCrNi, as well as SCI–VMn, is independent of the impact angle.

4. Discussion on eroded mechanism

4.1 Observation of wear feature of specimens

To make clear the erosive wear behaviors of selected specimens, the eroded surfaces were pictured by the digital camera after the erosion tests were finished. In the previous works of our group, in the case of employing steel grit, it was observed clear ripple pattern transverse to the impact direction of solid particles at 30°. Whereas at 60°, there is no clear ripple pattern, but indentation on upper part and ripple pattern just like at 30° from the center of eroded area to bottom. At 90°, there cannot be observed ripple pattern but indentation only [15,16]; that is to say, there are clearly different erosive wear mechanisms in shallow and high-impact angles that divided at about 60°. In the

131

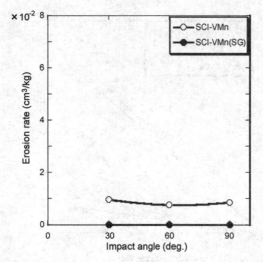

Fig. 4. Erosion rate vs. impact angle in SCI–VMn by steel grid and silica sand.

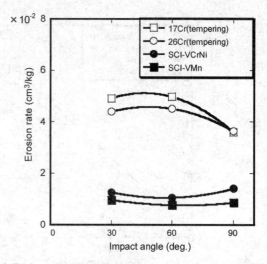

Fig. 5. Erosion rate vs. impact angle in specimens by silica sand particles.

present research, moreover, in the circumstance of silica sand, the macroscopic appearances of specimens have not shown any ripples. The features were shown in Fig. 6. It is observed that eroded area of materials are large and become longer along the impact direction at shallow angles of incidence and that the size of eroded surface areas are near round and eroded surface area became smaller at high angles of incidence in specimens. The reason why there have not formed ripples on the eroded surface is not become clear through macro-observation. Therefore, we

go into observation of the vertical section near the eroded surface by microscope. That will be discussed in the following parts.

4.2 Observation of vertical section near the eroded surface

In the case of employing steel grit, in the observation of vertical section near the surface, we also saw that there are formed tongue-shaped protruding portions that flowed along the impact direction plastically at 30° and 60°; whereas there were

132

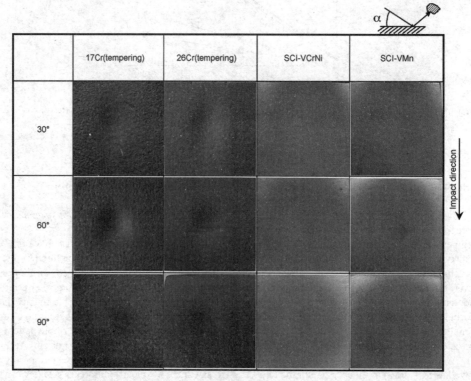

Fig. 6. Eroded surface of specimens by silica sand particles.

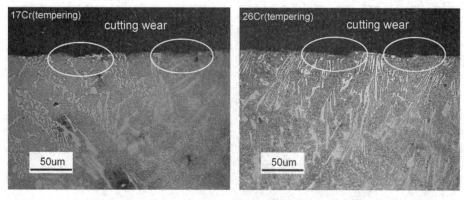

Fig. 7. Cross-section of 17Cr (tempering) and 26Cr (tempering) after erosion test at impact angles of 30°.

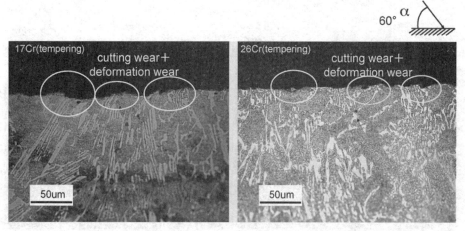

Fig. 8. Cross-section of 17Cr (tempering) and 26Cr (tempering) after erosion test at impact angles of 60°.

133

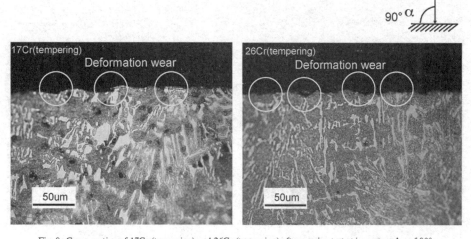

Fig. 9. Cross-section of 17Cr (tempering) and 26Cr (tempering) after erosion test at impact angles of 90°.

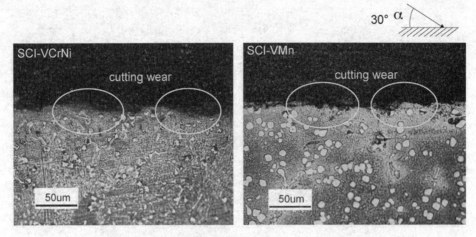

Fig. 10. Cross-section of SCI–VCrNi and SCI–VMn after erosion test at impact angles of 30°.

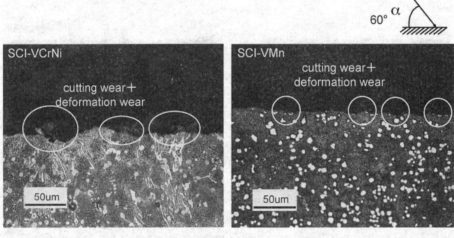

Fig. 11. Cross-section of SCI–VCrNi and SCI–VMn after erosion test at impact angles of 60°.

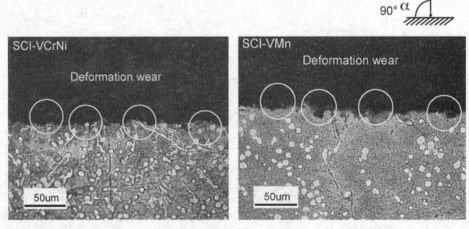

Fig. 12. Cross-section of SCI–VCrNi and SCI–VMn after erosion test at impact angles of 90°.

134

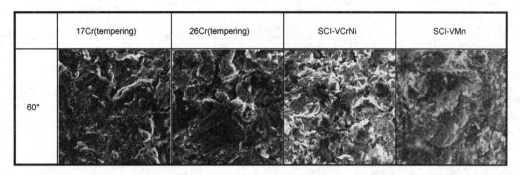

Fig. 13. SEM of eroded surface of specimens at 60° by silica sand particles.

formed the compressed piliform protruding portions and indentations resulting from plastic deformation at 90° in the previous researches [15,16]. On the other hand, in this work, employing silica sand, the vertical section near the surface is completely different from that using steel grit. The vertical section of WCIs showed in Figs. 7–9 in 30° and 60°, owing to cutting of silica sands, even the protruding portions were developed, they were smaller and brittle to brick away from test surface by impacting of following particles, at 90°, impacted from perpendicular direction, the test surface was broken and formed indentations on it, and then not only the indentations formed were expanded by following impact but also formed new indentations, these processes were repeated rapidly. Figs. 10–12 show the vertical section near surface of SCI–VCrNi and SCI–VMn with silica sand at 30°, 60°, and 90°, respectively. We observed from these figures that material was scraped away from surface by cutting of irregular shaped particles and formed protrusion, at the meantime, the spheroidal carbide was preserved in perfect shapes and the matrix surrounding it were scraped at the shallow angle of 30° and 60°. So we can understand from this fact that the spheroidal carbide is very hard and its adhesion on matrix is also wise. In the case of 90°, because of only compressive force stress on the material surface, it is confirmed that the spheroidal carbides not only get off from the eroded surface as it is but also break away from material leaving residual into the eroded surface.

And it is also seen clearly that the formed protruding portions were smaller than that formed in WCIs, although test surface extruded along the impact direction of particles, spheroidal VC adhere to matrix tightly which make test surface hard to plastic flow, and again spheroidal VCs are too hard to deform when impact by solid particles, matrix surrounded them fall off ahead. That is to say, because of the presence of spheroidal VC, the stress concentration that caused by impact was dispersed and that control the deformation of test surface, it can be suggested that the erosive wear resistances of SCIs are greater than WCIs.

135

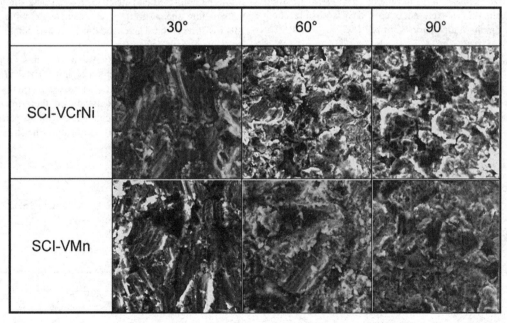

Fig. 14. SEM of eroded surface of SCI–VMn and SCI–VCrNi by silica sand particles.

4.3 SEM of specimens

SEM morphologies at impact angle of 60° for specimens are shown in Fig. 13; scrapes were evident on the eroded surface and at the end of the scrape there formed cracking lips, at the same time, big protruding portions and indentations were developed. This fact indicated that the erosive wear mechanism be likely to include what is called "cutting wear" and "deformation wear" that develop together during the erosion. SEM morphologies in Fig. 14 show the variation of erosive wear mechanism with the impact angle of particles. It can be clearly seen from the figure that at 30° the eroded surface was scraped off almost whole eroded area, on the other, the eroded surface is deformed plastically by impact particles to form big protruding portions and indentations in places at 90°. This will again indicate that, at the glancing angle, the "cutting wear" is predominant and, the "deformation wear" will be of greater importance with the increasing of impact angle of particles.

4.4 Vickers hardness of specimens

In the erosion test, it is recognized that the work-hardening effect which resulted from impact of solid particles on material surface was happened. The harnesses of specimens before and after erosion test were measured through Vickers hardness test.

Fig. 15 shows the results of them. From the differences of hardness before and after test, we can see that there were happened the work-hardening effects on the material surface. In SCI–VCrNi and SCI–VMn which both have superior wear-resistant properties, the results of Vickers hardness were dramatically increased after erosion test. The surface hardness of SCI–VCrNi increased from the initial 399 to 478 HV after 3600 s of blasting. Moreover, that of eroded SCI–VMn reached to 804 HV from the initial hardness of 533 HV.

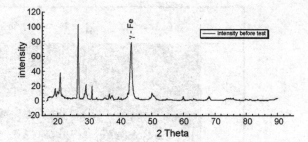

Fig. 16. X-ray diffraction spectrum of SCI–VMn before erosion test.

While although the initial hardness of 26Cr (tempering) and 17Cr (tempering) were higher than SCI–VCrNi and SCI–VMn, it almost not increased after erosion test and were more eroded than two kinds of spheroidal carbides cast irons. From this fact, we can understand that it is true that erosion rate depends heavily on hardness of surface of material [2], however, compare with the initial hardness of material, the hardness after work hardened is likely to be of more importance.

4.5 X-ray diffraction analysis on eroded surface

Figs. 16 and 17 illustrate the XRD patterns for both initial and eroded surface on SCI–VMn. Fig. 16 shows that before erosion test austenite matrix is dominant phase and after test (Fig. 17) austenite has decrease to 40% and martensite phase has occurred. From the result above, it suggests that when particles impact on the surface the austenite in matrix in SCI–VMn change to martensite by strain-induced transformation, which results in increasing of hardness and sound erosive wear resistance. And then, with the increasing of impact time, formed martensite would be deformed and stripped from surface, simultaneously, the structure under surface again changes to martensite repeatedly to form new hard layer and this layer maintains certain thick from the eroded surface.

Finally, the relationship between materials properties and wear resistances of specimens are summarized in Table 3. It can be understood that at the same wear condition, erosive wear of materials heavily depends on their microstructural properties and mechanical characters. So we have to allow for material designs to be applicable in strict industrial conditions.

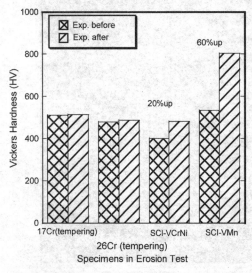

Fig. 15. Change of Vickers hardness for specimens after erosion test.

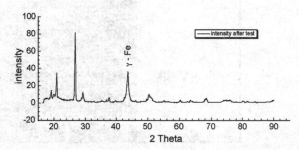

Fig. 17. X-ray diffraction spectrum of SCI–VMn after erosion test.

Table 3
Summaries of characteristics of alloys

Material	Microstructural features (60°)	Hardness changes (HV)	Erosion rate ($\times 10^{-2}$ cm^3/kg)	Wear mechanism
17Cr (tempering)		$510 \rightarrow 514$	30°: 4.9143, 60°: 4.9702, 90°: 3.5858	The small and brittle protruding portions were developed and brick away from test surface Indentations formed and expanded These processes were repeated rapidly
26Cr (tempering)		$478 \rightarrow 485$	30°: 4.3959, 60°: 4.5122, 90°: 3.6370	The small and brittle protruding portions were developed and brick away from test surface Indentations formed and expanded These processes were repeated rapidly
SCI–VCrNi		$399 \rightarrow 482$	30°: 1.2595, 60°: 1.0491, 90°: 1.4065	Spheroidal VCs make test surface hard to plastic flow, and again are too hard to deform, matrix surrounded them fall off ahead The presence of spheroidal VC dispersed the stress concentration
SCI–VMn		$533 \rightarrow 804$	30°: 0.96020, 60°: 0.76080, 90°: 0.85510	Spheroidal VCs make test surface hard to plastic flow, and again are too hard to deform, matrix surrounded them fall off ahead The presence of spheroidal VC dispersed the stress concentration

137

5. Conclusions

Erosion tests were carried out on two kinds of spheroidal carbides cast irons and two kinds of high chromium cast irons with irregular shaped sand to investigate the erosion behaviors and mechanisms of spheroidal carbides cast iron. The results showed clearly that erosion rates differed with structures and types of materials and there are different kinds of erosion rate and impact angle dependency. And then the following conclusions were obtained:

(1) Cast iron with spheroidal carbides has clearly superior erosive wear resistance compare with high chromium cast irons. The reasons maybe that stress concentration was dispersed because carbides in matrix structure are spheroidized through adding vanadium.

(2) SCI–VMn, as well as SCI–VCrNi has higher erosive wear resistance than high chromium cast irons and SCI–VMn has most superior erosive wear resistance among the materials employed, owing to obvious work-hardening effect.

(3) In erosive wear, the hardness after work hardening is of more importance than the initial hardness. And the strain-induced transformation effect of matrix in SCI–VMn has direct effect on increasing erosive wear resistance.

(4) Impact angle is one of the major controlling factors of erosive wear. Here, erosion tests were performed with impact

angles of 30°, 60° and 90° to determine the relationship between impact angle and erosion rate. However, erosion rates of SCI–VMn and SCI–VCrNi are independent on the impact angle of particles, that is to say, SCI–VMn and SCI–VCrNi is ideal materials to use as the bended section of the pipe in the strict circumstances.

(5) The fact that erosive wear resistance will increase when spheroidal carbide is precipitated in structure of specimens was made clear. It is considered that if carbides were precipitated with spheroidal shapes, the formed stress concentration, which will cause the forming of protuberance, would be dispersed, and then it will result in the increasing of the property of erosive wear resistance of spheroidal carbides cast irons.

Acknowledgements

The authors would like to thank the other staffs of Material Processing Study Room at Muroran Institute of Technology Japan for giving assistances during all experimental works.

References

[1] I. Finnie, Erosion of surfaces by solid particles, Wear 3 (1960) 87–103.

[2] J.G.A. Bitter, A study of erosion phenomena. Part I, Wear 6 (1963) 5–21.

[3] J.G.A. Bitter, A study of erosion phenomena. Part II, Wear 6 (1963) 169.

[4] J.H. Neilson, A. Gilchrist, Wear 11 (1968) 111.

[5] G.P. Tilly, Erosion caused airborne particles, Wear 14 (1969) 63–79.

[6] G.P. Tilly, Sand erosion of metals and plastics: a brief review, Wear 14 (1969) 241–248.

[7] G.P. Tilly, W. Sage, The interaction of particle and material behaviour in erosion processes, Wear 16 (1970) 447–465.

[8] G.P. Tilly, A two stage mechanism of ductile erosion, Wear 23 (1973) 87–96.

[9] I.M. Hutchings, R.E. Winter, Particles erosion of ductile metals: a mechanism of material removal, Wear 27 (1974) 121–128.

[10] I.M. Hutchings, Mechanism of the Erosion of Metals by Solid Particles, ASTM STP664, 1979, p. 59–76.

[11] I. Hussainova, Microstructure and erosive wear in ceramic-based composites, Wear 258 (2005) 357–365.

[12] D.W. Wheeler, R.J.K. Wood, Erosion of hard surface coatings for use in offshore gate valves, Wear 258 (2005) 526–536.

[13] M. Divakar, V.K. Agarwal, S.N. Singh, Effect of the material surface hardness on the erosion of AISI316, Wear 259 (2005) 110–117.

[14] S.B. Mishra, S. Prakash, K. Chandra, Studies on erosion behaviour of plasma sprayed coatings on a Ni-based superalloy, Wear 260 (2006) 422–432.

[15] Kazumichi Shimizu, Toru Noguchi, Erosion characteristics of ductile iron with various matrix structures, Trans. Jpn. Foundrymen's Soc. 13 (1994), NOV.

[16] Kazumichi Shimizu, Toru Noguchi, Fundamental study on erosive wear of austempered ductile iron, in: Proceeding of the 3rd East Asian International Foundry Symposium, Pusan, Korea, 1–3 July, 1992.

[17] K. Shimizu, T. Noguchi, S. Doi, Basic study on the erosive wear of austempered ductile iron, Trans. AFS 101 (1993) 225–229 (paper nos. 93–78).

[18] K. Shimizu, T. Noguchi, T. Kamada, H. Takasaki, Progress of erosive wear in spheroidal graphite cast iron, Wear 198 (1996) 150–155.

[19] K. Shimizu, T. Noguchi, T. Kamada, S. Doi, Basic study of erosion of ductile iron, Adv. Mater. Res. 4–5 (1997) 239–244.

138

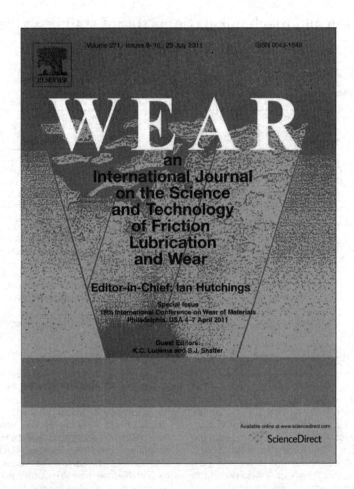

Volume 271, Issues 9-10, 29 July 2011   ISSN 0043-1648

WEAR

an
International Journal
on the Science
and Technology
of Friction
Lubrication
and Wear

Editor-in-Chief: Ian Hutchings

Special Issue
18th International Conference on Wear of Materials
Philadelphia, USA 4-7 April 2011

Guest Editors:
K.C. Ludema and S.J. Shaffer

Available online at www.sciencedirect.com
ScienceDirect

139

Contents lists available at ScienceDirect

Wear

journal homepage: www.elsevier.com/locate/wear

Solid particle erosion and mechanical properties of stainless steels at elevated temperature

K. Shimizu*, Y. Xinba, S. Araya

Muroran Institute of Technology, 27-1, Mizumoto-cho, Muroran city, Hokkaido, 050-8585, Japan

ARTICLE INFO

Article history:
Received 1 September 2010
Received in revised form
22 December 2010
Accepted 22 December 2010

Keywords:
Mechanical property
High temperature erosion
High temperature hardness
Elongation

ABSTRACT

The mechanical properties of a target material greatly influence its erosion behaviour at elevated temperature. This study investigated the correlation between the mechanical properties of stainless steels at high temperature and its resistance to erosion to estimate high temperature erosion behaviour. Two types of stainless steels, SUS403 and SUS630, were selected. High temperature solid particle erosion tests were performed using 1 mm alumina particles, at impact angles between 30° and 90° and a particle velocity with 100 m/s, with changing test temperature of room temperature, 573 K, 873 K and 1173 K.

To clarify the correlation of the mechanical properties and the erosion resistance of the stainless steels, high temperature hardness and tensile tests were also carried out for SUS403 and SUS630, to estimate the erosion damage caused by solid particles. The erosion at an impact angle of 90° for both SUS403 and SUS630, where deformation wear is predominant, suggested that there is a proportional relationship between the decreasing percentage of hardness and the erosion rate. In the case of an impact angle of 30° and test temperature of 1173 K, the erosion rate of SUS403 was twice that of SUS630, suggesting that at a shallow angle, material with a higher elongation easily forms protrusion that can be removed by impacting.

1. Introduction

Erosion is caused by a gas or a liquid which may or may not carry entrained solid particles, impinging on a surface. For an instance, during the manufacturing of heat insulator, a high temperature slag impacts on a high speedy rotating rotor, and then spread out with flocculating style by the centrifugal force of the rotating rotor, on the surface of which erosion damage occur to become a big industrial problem. Similarly, instances of sand erosions occur in secondary refining and smelting reduction furnace at iron and steel plants. When the dispersed particles such as dust coals, powdered minerals are blown into the melted pig iron through blow pipes, the erosion damage always occur at the bended sections of the pipes.

The mechanism of erosion by solid particles is very complex. Erosion factors are the mechanical properties of target materials, kinetic energy, mass, hardness and shapes of the erodent particles, and then impact speed and impact angle of particles, etc. [1–4]. Among them, the dominant factors that influence erosion are mainly associated with the mechanical properties of target materials.

Many researchers have been focused on the mechanical properties of target materials. Gat and Tabakoff [5] have found that the erosion damage depends upon the impact angle and upon the test temperature with respect to the thermal properties of the material. Naim and Bahadur [6] have noted the erosion rate of some heat treated steels depend on both the hardness and the ductility at ambient condition. Levy et al. [7] have investigated the relationship between erosion rate and temperature, to emphasis that there have marked increase in erosion rate over certain temperature and hardness had no correlation with their erosion rate.

Moreover, it has long been recognized that during the erosion process, the maximum erosion rate occurs at shallow angle of 20–30° for ductile materials (aluminum and copper) [1–3], and brittle materials (ceramics and glass) reaches highest erosion rate approximately at higher impact angles of 80–90° [8–10]. What is more, the environmental factors also a great effect on the mechanical performance of materials [11], assuming that as test temperature increased, the hardness and tensile strength of materials are changed, which may influence the erosion characteristics of materials. Although these previous studies have provided a lot of meaningful understanding, detailed information about the relationship between materials features and the erosion behaviours in the high temperature region have not yet completely recognized.

To clarify the correlation of the mechanical properties and the erosion resistance of the material could contribute to the solution

---

* Corresponding author. Tel.: +81 143 465651; fax: +81 143 465651.
E-mail address: shimizu@mmm.muroran-it.ac.jp (K. Shimizu).

0043-1648/$ – see front matter © 2011 Elsevier B.V. All rights reserved.
doi:10.1016/j.wear.2010.12.038

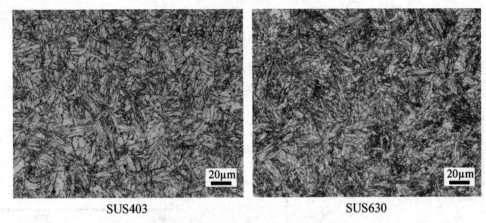

SUS403   SUS630

Fig. 1. Microstructure of two types of stainless steels.

of the erosion problem not only at room temperature but also at high temperature region in the steel plants.

This study investigated the effect of mechanical properties of the stainless steels on their erosion rates as a function of impact angle of the erodent at high temperature. Emphasis was laid in the relationships among the mechanical properties (hardness and elongation), its erosion rates and impact angles of erodent at high temperature.

High temperature erosion tests, high temperature hardness and tensile tests are carried out for two types of stainless steels of SUS403 and SUS630. How the mechanical properties effect on the erosion resistance of target materials were investigated and the propriety of the presumption of high temperature erosion resistance behaviour from their mechanical properties at high temperature was argued.

2. Experimental procedures

2.1 High temperature erosion tests

High temperature erosion tests were carried out using a newly developed high temperature erosion setup. Specimens were mounted into a test stage, and then were heated to 1173 K in the heating chamber. 800 g erodent were loaded and were heated to 1073 K in the tank. During the test, the temperatures of specimen, erodent and hot air was monitored by a control board that connected with thermocouples attached to the specimen, erodent and hot air generator. The testing atmosphere is controlled to constant. A detailed description of the high temperature erosion setup was given in Ref. [11]. Erosion test duration was 300 s and the amount of erodent was 8 kg/test. Impact feed was 26 g/s and impact speed was 100 m/s. Testing conditions were summarized in Table 1.

The weight loss of a specimen was determined using an electronic scale. The erosion rate was then defined as following

Table 1
High temperature erosion test conditions.

Test variable	Test conditions
Erodent particle temperature T (K)	1073, 773, room temperature
Test piece temperature T (K)	1173, 873, room temperature
Entrained air temperature T (K)	773, room temperature
Hot air velocity V (m/s)	100 m/s
Impact angle a (°)	30, 60, 90
Test duration t (s)	300
Erodent particle loading L (kg)	8
Target materials	SUS403, SUS630
Erodent material	Alumina ball (1 mmϕ)

formulas to evaluate the removed material volume.

Volumetric removal per second (cm^3/s)

$$= \frac{\text{Mass removal per second (g/s)}}{\text{Average density (g/cm}^3)}$$

Erosion rate (cm^3/kg)

$$= \frac{\text{Volumetric removal per second (cm}^3\text{/s)}}{\text{Mass amounts of impact particles per second (kg/s)}}$$

The reason why was the volumetric removal divide by mass amounts of impact particles is that the mass removal per second should depends on the total amounts of particles that impact on target material in that period. It was considered that the more the amounts of particles impact on surface of material the more the material removal increased. So it is more reasonable to determine the value at amount of impact particles per second as erosion rate.

2.2 Materials

Materials selected for the present study were two types of stainless steels, SUS403 and SUS630. The chemical compositions and Vickers hardness of the alloys were given in Table 2. The size of

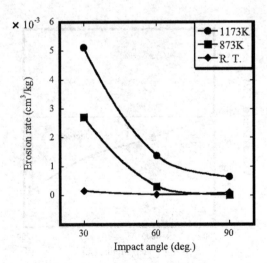

Fig. 2. Erosion rate as a function of impact angle at variety of test temperature for SUS403.

141

Table 2
Chemical compositions and hardness of alloys.

	C	Ni	Cr	Cu	Nb	Vickers hardness
SUS403	0.11	0.40	12.18	–	–	193
SUS630	0.05	3.81	16.27	3.65	0.34	338

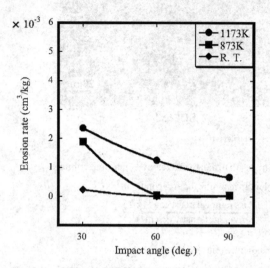

Fig. 3. Erosion rate as a function of impact angle at variety of test temperature for SUS630.

a specimen was 50 mm × 50 mm × 10 mm. The test surface was ground with surface finish 0.2 μm Ra. The metallographic structures of two materials were illustrated in Fig. 1.

2.3 High temperature Vickers hardness test

The hardness test was conducted using AVK-HF type high temperature hardness tester. Specimen was measured to 6.5 mm × 6.5 mm × 5.0 mm, and test surface was finished with polishing process. Test temperatures were room temperature, 573 K, 873 K and 1173 K. The temperature error range was assumed to be ±5 °C. Temperature increasing speed was 10 °C/min. Hardness test

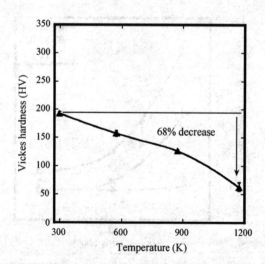

Fig. 4. Vickers hardness as a function of testing temperature for SUS403.

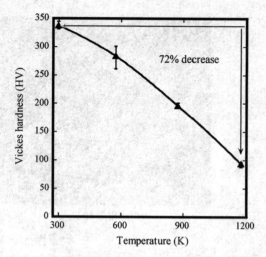

Fig. 5. Vickers hardness as a function of testing temperature for SUS630.

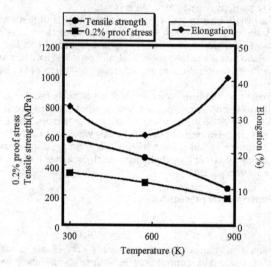

Fig. 6. Results of high temperature tensile test for SUS403 (tensile strength, 0.2% proof stress and elongation).

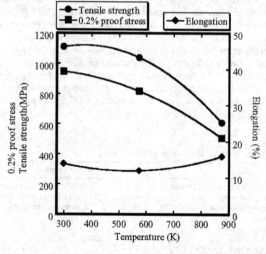

Fig. 7. Results of high temperature tensile test for SUS630 (tensile strength, 0.2% proof stress and elongation).

142

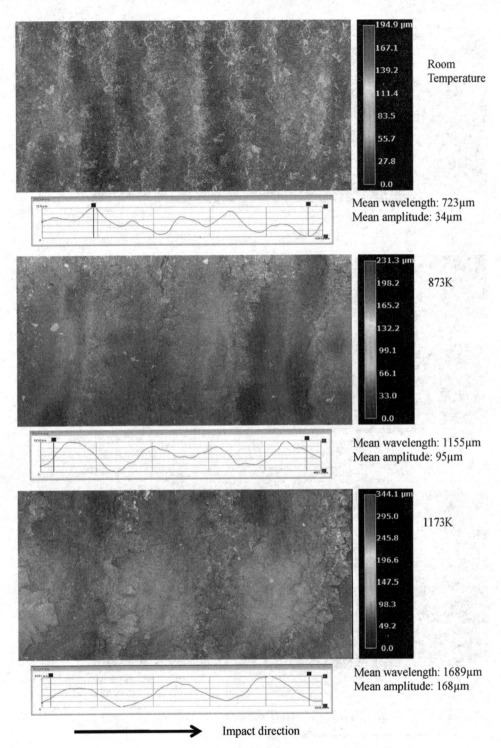

Mean wavelength: 723μm
Mean amplitude: 34μm

Mean wavelength: 1155μm
Mean amplitude: 95μm

Mean wavelength: 1689μm
Mean amplitude: 168μm

Impact direction

Fig. 8. Surface morphology of SUS403 after erosion test at impact angle of 30°.

was conducted after keeping on the certain test temperature for 5 min. Argon (or Ar) was blown into test chamber to control the oxidation of test surface. The diamond indenter was used at a load of 1 kgf with 10 s loading. The average values of 10 reading points were noted.

2.4 Tensile test

The tensile strength, which is also an important factor that influences the erosion of material, differs with test temperature. The changes in erosion rate at high temperature can

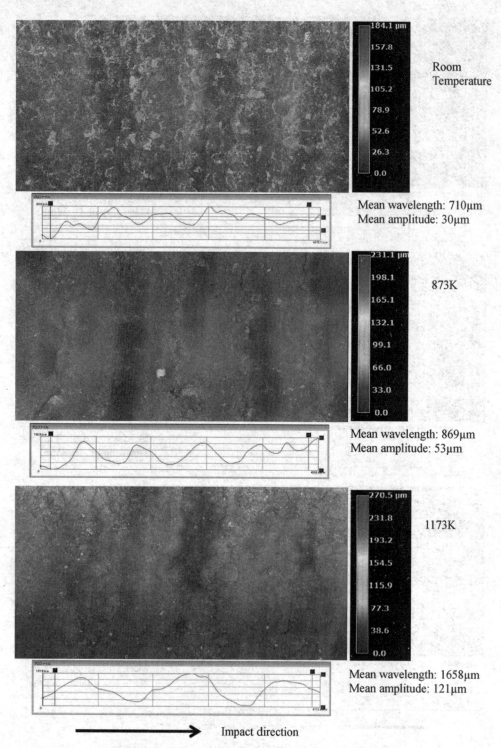

Room
Temperature

Mean wavelength: 710μm
Mean amplitude: 30μm

873K

Mean wavelength: 869μm
Mean amplitude: 53μm

1173K

Mean wavelength: 1658μm
Mean amplitude: 121μm

Impact direction

Fig. 9. Surface morphology of SUS630 after erosion test at impact angle of 30°.

144

be related to changes of high temperature tensile strength of alloys [7]. In this study, tensile tests at deferent temperature were performed to investigate its correlation to the erosion behaviour. Specimens were prepared according to the JIS Z2201 14A, with changing test temperature of room temperature, 573 K and 873 K, respectively. During all of tests, the speci-mens were maintained at the mentioned temperature within ±3 °C. An average value was obtained from 3 tests carried out.

In addition, digital microscope and scan electron microscope (SEM) observations were carried out to characterize the erosion mechanism for both metals.

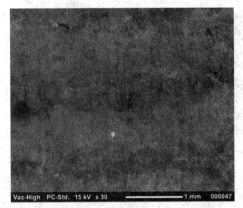

SEM micrographs showing the ripple patterns formed at impact angle of 60° for 1173K in SUS403.

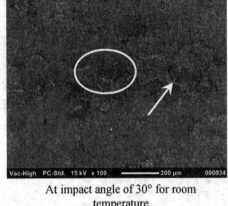

At impact angle of 30° for room temperature

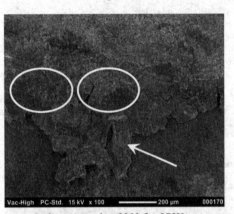

At impact angle of 30° for 873K

SEM micrographs showing the random indentations but no ripples formed at impact angle of 90° for1173K in SUS403.

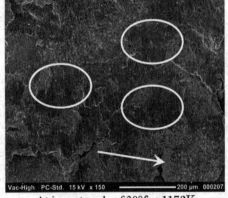

At impact angle of 30°for 1173K

Fig. 10. SEM micrographs of SUS403 after erosion test.

3. Results and discussions

3.1 *High temperature erosion rate*

Relationships between erosion rates and impact angles of erodent for SUS403 and SUS630 are shown in Figs. 2 and 3 respectively. For each specimen, erosion rates tent to increase with testing temperature. And for all of testing temperatures, erosion rates reached maximum values at an impact angle of 30° and followed by a smooth decrease to lowest values at 90°. The specimens displayed a typical impact angle dependency that was shown for ductile material [12]. At same temperature and same impact angle, the erosion rates of SUS403 were higher than that of SUS630, especially, at 1173 K for 30°, the value of SUS403 was twice higher than that of SUS630.

3.2 *High temperature Vickers hardness tests*

The hardness test results of SUS403 and SUS630 are shown in Figs. 4 and 5. The figures revealed that the hardness of two

145

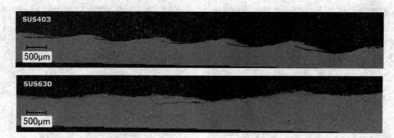

Fig. 11. Cross section profiles of near surface for SUS403 and SUS630 at temperature of 1173 K and impingement angle of 30°.

materials is dramatically decreased with temperature. However, at all of temperatures, SUS630 is harder than SUS403 by approximately 1.8 times at room temperature and 573 K, approximately 1.5 times at 873 K and 1173 K.

3.3 High temperature tensile tests

Figs. 6 and 7 show the results of tensile tests for SUS403 and SUS630. Comparing Figs. 6 and 7, 0.2% proof stress and tensile strength of SUS630 are twice higher than that of SUS403 at room temperature, and the differences are exceeded to 2.5 times when testing temperature is elevated to 873 K. On the other hand, the elongation of SUS630 is 2.4 times less than that of SUS403. The difference of elongations of two steels is also extended to 2.6 times at 873 K.

3.4 Relationship between mechanical properties and erosion at high temperature

Surface morphologies of SUS403 and SUS630 with increasing test temperature are determined using VHX-100 series digital microscope (a product of Keyence) and SEM. The surface morphologies of SUS403 and SUS630 at impact angle of 30° pictured by digital microscope are shown in Figs. 8 and 9. The figures illustrate well-developed ripple patterns lying transverse to impact direction of erodent for both materials. A typical profile of ripple pattern for each test temperature is illustrated below each graph. The mean wavelength and amplitude of the ripples were determined by test temperature, i.e., the higher the test temperature the longer the wavelength of ripples with increasing of their mean amplitude. It indicated that when the hardness of materials was decreased with test temperature (as shown in Figs. 4 and 5), the materials showed ductile characters. Consequently, the erosion rates were increased with increasing of test temperature for both SUS403 and SUS630. Moreover, compare the both materials the wavelength and amplitude of SUS630 at each test temperature were smaller than that of SUS403. Considering relative to hardness values at each test temperature for SUS403 and SUS630, it was not difficult to understand the fact that erosion rate of SUS630 are lower than that of SUS403 at each temperature.

As will be illustrated shortly by SEM observation there are also confirmed similar ripple patterns at impact angle of 60°, while for impact angle of 90°, there are developed random indentations but with no coherent ripple patterns at high test temperature of 1173 K.

As shown in Figs. 2 and 3, compare to SUS630, erosion rate of SUS430 is twice higher than that of SUS630 at high temperature of 1173 K with the impact angle of 30°. The reason was likely to relate to hardness and elongation at elevated temperature. According to Figs. 4–7, high temperature hardness of SUS630 is higher than that of SUS403, while elongation of SUS630 is lower than that of SUS403 at each testing temperature. Accordingly, it considered that when impacted by particle, although the surface was deformed plastically and formed big protrusion on test surface, the progress

on SUS630 was less apparent than that on SUS403. Moreover, the surface structure of SUS630 was hardly protruded by subsequent impact so that the erosion of SUS630 was restrained. On the other hand, from Figs. 6 and 7, it was clearly observed that the elongation of SUS630 is not very changed with temperature while that of SUS403 is changed greatly. The smaller changing ratio of elongation of SUS630 than that of SUS403 with temperature was considered to contribute to the smaller impact angle dependency of SUS630.

In particular, as for both materials, the high temperature hardness changing ratios from room temperature to 1173 K did not have so many differences (68% decrease for SUS403 and 72% for SUS630). While, in the case of test temperature of 1173 K at impact angle of 30°, the elongations of SUS403 showed twice higher than that of SUS630. Because erosion at 30° progresses by formation, extension of the protrusion followed by its falling off from test surface, the influence of the elongation became remarkable. Consequently, the protrusions formed for SUS630 were not extended by impact of erodent particles, and then wear loss was decreased.

On the other hand, Figs. 2 and 3 revealed that the increases of erosion rates at 90° with testing temperature for two materials are similar when testing temperature is increased from the room temperature to 1173 K. The reasons were considered that at the impact angle of 90° where deformation wear is predominant for erosion in general, the decreases on hardness allow test surfaces to display ductile behaviour and then easy to deform plastically. The result indicated that erosion rate increase is directly proportional to hardness decreasing ratio at high impact angle of 90° when testing temperature is elevated.

As illustrated in Fig. 10, the SEM micrographs of the eroded surface show the pilled up protrusions at the crests of ripple patterns at impact angle of 30° in detail for room temperature, 873 K and 1173 K, respectively. These revealed that the formed protrusions at room temperature are small and with increasing of temperature, they are grown to be large as pointed with arrows in figures. These graphs also revealed that the crests of the ripples are cut by erodent and these cutting traces are increased with temperature. The evident are shown by the circles in the figures.

Similar evident are shown in the eroded surface of SUS630. But the length of formed protrusions appeared to be shorter than that of SUS403. Fig. 11 shows cross section observation of SUS403 and SUS630, and Table 3 shows the length of protrusion formed on the crests of ripples. Protrusions of SUS403 were extended largely and the average value was approximately 530 μm. On the other hand, the average length of protrusion of SUS630 was approximately 266 μm, compare to that of SUS403, which was not progressed so much. This fact also indicated that extension of protrusion was

Table 3
Protrusion length of protruding portions.

	Maximum (μm)	Minimum (μm)	Average (μm)
SUS403	885	216	530
SUS630	457	117	266

greatly influenced by the elongation of material. High temperature erosion behaviour of material, where erosion rate at impact angle of 30° get maximum value, was dominated by the value of erosion rate at 30°. Therefore, the smaller the erosion rates at 30° for materials, the smaller the impact angle dependency of them. This leads to a conclusion that impact angle dependency is greatly affected by elongation at elevated temperature for materials.

4. Conclusions

The aim of this study was to clarify the correlation of the mechanical properties and the erosion behaviour of the stainless steels at high temperature. Two types of stainless steels, SUS403 and SUS630, were employed for the high temperature erosion tests and the high temperature hardness and tensile tests. The following results are obtained.

1) In the case of high temperature erosion, the material softens by the high temperature, and the cutting wear is appeared remarkably.
2) The erosion at an impact angle of 90°, where deformation wear is predominant, suggested that there is a proportional relationship between the decreasing percentage of hardness and the erosion rate.
3) In the case of impact angle of 30° at a test temperature of 1173 K, the reason there appeared twice the difference of erosion rate

between materials is thought to be a difference of the elongation, suggesting that in the case of erosive wear at a shallow angle, the material with a higher elongation easily formed protrusion that can be removed by successive impacting of erodent.

References

[1] I. Finnie, Erosion of surfaces by solid particles, Wear 3 (1960) 87–103.
[2] K. Shimizu, T. Noguchi, T. Kamada, H. Takasaki, Formation and progression of erosion surface in spheroidal graphite cast iron, Trans. AFS 104 (96–61) (1996) 511–515.
[3] K. Shimizu, T. Noguchi, Erosion characteristics of ductile iron with various matrix structures, Wear 176 (1994) 255–260.
[4] X. Yaer, K. Shimizu, et al., Erosive wear characteristics of spheroidal carbides cast iron, Wear 264 (2008) 947–957.
[5] N. Gat, W. Tabakoff, Some effects of temperature on the erosion of metals, Wear 50 (1978) 85–94.
[6] M. Naim, S. Bahadur, Effect of microstructure and mechanical properties on the erosion of 18 Ni (250) maraging steel, Wear 112 (1986) 217–234.
[7] A.V. Levy, J. Yan, J. Patterson, Elevated temperature erosion of steels, Wear 108 (1986) 43–60.
[8] I. Finnie, J. Wolak, Y. Kabil, Erosion of metals by solid particles, J. Mater. 2 (1967) 682–700.
[9] Y.I. Oka, H. Ohnogi, T. Hosohawa, M. Matsumura, The impact angle dependence of erosion damage caused by solid particle impact, Wear 203–204 (1997) 573–579.
[10] R.G. Wellman, C. Allen, The effects of angle of impact and material properties on the erosion rates of ceramics, Wear 186–187 (1995) 117–122.
[11] K. Shimizu, T. Naruse, Y. Xinba, K. Kimura, K. Minami, H. Matsumoto, Erosive wear properties of high V–Cr–Ni stainless spheroidal carbides cast iron at high temperature, Wear 267 (2009) 104–109.
[12] I. Finnie, Some observations on the erosion of ductile metals, Wear 19 (1972) 81–90.

Contents lists available at ScienceDirect

Wear

journal homepage: www.elsevier.com/locate/wear

High temperature erosion characteristics of surface treated SUS410 stainless steel

K. Shimizu [a,*], Y. Xinba [a], M. Ishida [a], T. Kato [b]

[a] Muroran Institute of Technology, 27-1 Mizumoto-cho, Muroran city, Hokkaido, 050-8585, Japan
[b] Trytec. Co. Ltd, 335 Kuwahara Kannami-cho, Shizuoka, 419-0101, Japan

ARTICLE INFO

Article history:
Received 1 September 2010
Received in revised form 6 January 2011
Accepted 7 January 2011

Keywords:
Erosion
High temperature
High temperature hardness
Surface treated
Plastic flow

ABSTRACT

This study investigated the high temperature erosion characteristics of two types of surface-treated SUS410 steels; overlay welding and forging of the base metal. Two-layer overlay welding of 6 mm and forging with a 10% reduction, were used on a base metal of SUS410, to prepare specimens. High temperature solid particle erosion tests using a test temperature of 1173 K were performed using 1 mm alumina particles, with impact angles between 30° and 90° and a particle velocity of 100 m/s. Erosion rates, especially at shallow angles of 30°, were dramatically different for all specimens. Compared with the base metal of SUS410, the erosion rate was reduced by 50% for overlay welded material, and 30% for forged material.

High temperature hardness measurement and the observation of the eroded surface by scan electron microscopy were undertaken to analyse the erosion behaviour. An increase in the erosion rate of the specimen was related to a decrease in the high temperature hardness. Although the hardness was reduced to approximately 70% at 1173 K for all specimens, this suggested that the wear resistance of the overlay welding material was improved by restraining the plastic flow because it was harder at high temperatures. The forged material was suggesting that the plastic flow of eroded surface was restrained by the refinement of the microstructure and the residual stress near the surface, which reduced erosion rate regardless of this lower hardness.

Published by Elsevier B.V.

149

1. Introduction

Erosive wear is caused by a gas or a liquid which may or may not carry entrained solid particles, impinging on a surface [1,4]. High temperature erosion is one of the main failures of industrial plants. An instance of high temperature erosion occurs in the production of the inorganic fibrous insulator in plant. In the manufacturing process, the melting slag impacts on a highly rotating rotor and is stretched out by the centrifugal force of this rotor. During the procedure, the outer surface of rotor is eroded by impact of the melted slag. Fig. 1 illustrates the model of the rotors and the erosive wear appearance of a rotor which is occurred at high temperature of 1073 K or over. As revealed in Fig. 1, the melted slag (approximately 1773 K) is poured in to a rotated rotor 1 and then impact on a rotated rotor 2 with a decrease of temperature to approximately 1073 K, where the parts of slag is solidified due to the decreased temperature. Rotor 2 is eroded by the impact of the melted partially solid slag. The main components of the slag used in the plant were shown in Table 1. To improve the yield rate and the production quality of the inorganic fibrous insulator, increasing of the centrifu-

gal force of rotor is an effective way. When the number of revolution of rotor is increased, however, the service life of rotor decreases to 1/10–1/2 of the former one, which makes the increasing of rotor exchange frequency and costs. The rotor materials (stainless steel) used up to date, are not the materials which could be coping with wear resistance and thermal resistance. The development of new materials which both has wear resistance and thermal resistance is on the great demand.

Erosion factors are kinetic energy, mass, hardness and shapes of erodent particles, etc. Amount of material removal of ductile materials appear to be at maximum value in the range of 20°–30°, on the other hand, that of brittle materials seem likely to be at maximum value in the range of 80°–90° [1–3]. That is to say, it is generally that material removal is dependent on the impact angle of particles and there have weak angle and strong angle in a material.

It has been cleared that the wear removal increases when temperature elevated, and mostly the steel material shows maximum value at the shallow angle that is the feature of ductile materials for high temperature erosion [4–9]. However, there are a lot of uncertainties that have to be clarified in the high temperature erosion.

The surface modification has received considerable interest because it can minimize the erosion damage that occurred not only at room but also high temperature region [10–16]. Accordingly, the

* Corresponding author. Tel.: +81 143 465651; fax: +81 143 465651.
E-mail address: shimizu@mmm.muroran-it.ac.jp (K. Shimizu).

0043-1648/$ – see front matter. Published by Elsevier B.V.
doi:10.1016/j.wear.2011.01.055

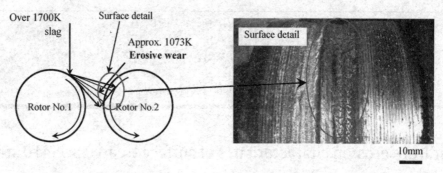

Fig. 1. A model of the inorganic fibrous insulator machine rotors and the eroded surface of rotor.

Table 1
The main components of the melting slag.

WT%					
SiO_2	Al_2O_3	Fe_2O_3	MgO	CaO	MnO
0.35–0.45	0.10–0.20	0–0.03	0.04–0.08	0.30–0.40	0–0.01

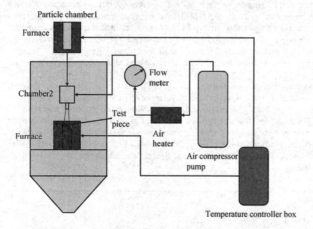

Fig. 2. The schematic diagram of the high temperature erosion setup.

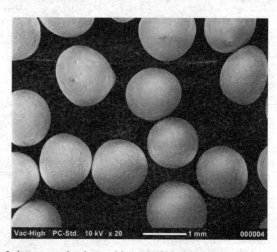

Fig. 3. Appearance of erodent particles used in high temperature erosion setup.

present study is paid attention to the surface modified material as a countermeasure of the erosion in the high temperature region, and tempts to clarify its high temperature erosion mechanism.

2. Experimental procedure

2.1 High temperature erosion tests

In this study, to understand the erosive wear in the region of high temperature, new high temperature erosion test machine was manufactured aiming to simulate the condition of the manufacturing process of the inorganic fibrous insulator plant. The high temperature erosion test machine has been introduced in details in earlier works [17]. The schematic diagram of the erosion setup is illustrated in Fig. 2. The machine consists of two furnaces (heating chambers) where test piece and erodent are heated respectively and a separate heater where compressed air is heated. Particles are supplied from the particle chamber 1 into the chamber 2 where entrained by hot air to fly out from the nozzle.

The erodent particle used in this study was a 1 mm alumina ball. The particle appearance of this erodent was shown in Fig. 3 and some characteristics of it were listed in Table 2. The erodent particles were refreshed after each erosion test. Specimens with dimensions $50 \times 50 \times 10$ (mm) were used for the high tempera-

ture erosion test. The specimens were mounted into the test stage directly below the nozzle with a vertical distance of 50 mm from the end of the nozzle to the test surface into the erosion test machine by changing their impact angles 30°, 60° and 90°, respectively. The examined air speed was 100 m/s, and the particle feed rate was approximately 26 g/s. Table 3 showed the experimental conditions.

2.2 Materials

In this study, the forging process and the overlay welding were performed as a surface modification method for SUS410 base material. SUS410 was quenched at 1223 K and tempered at 1023 K. As for the forged SUS410, after softening by annealing, SUS410 was forged with a 10% reduction using cold forging process. While, The two-layer overlay welding method was used to produce overlay welded

Table 2
Some characteristics of erodent particle.

Erodent particle	Alumina ball
Specific gravity	3.6 or greater
Vickers hardness	1100
Components	Alumina (Al_2O_3) purity 92% or greater
Average size (mmϕ)	1

150

Table 3
High temperature erosion test conditions.

Test variable	Test conditions
Erodent particle temperature T (K)	1073, room temperature
Test piece temperature T (K)	1173, room temperature
Entrained air temperature T (K)	773, room temperature
Hot air velocity V (m/s)	100 m/s
Impact angle a (°)	30, 60, 90
Test duration t (s)	300
Erodent particle loading L (kg)	8
Target materials	SUS410, forged SUS410, overlay welding SUS410
Erodent particle	Alumina ball (1 mm ϕ)

material; the joint layer was of soft property and good deposit to the base metal, and the test surface layer was of hard property, welding electrode with main component of chromium carbides was used to make this layer. The specimen was sized with $50 \times 50 \times 10$ (mm), and each overlay thickness was 3 mm respectively. And base material of SUS410 was prepared as a comparison material.

Fig. 4 showed the microstructure of SUS410, forged SUS410 and overlay welded SUS410. SUS410 and forged SUS410 were etched by an alcohol liquid composed of 15% hydrochloric acid and 1% picric acid. Forged SUS410 was compressed to form grain structure overall. Nital was used to etch the overlay welded material. The variety sized carbides appeared in the matrix of second layer. In order to determine the composition and distribution of the alloying elements, EPMA surface analysis was performed at the surface of overlay welding material (the hard layer) using EPMA (a product of JEOL). The result was shown in Fig. 5. EPMA surface analysis revealed that C, Cr, and Si were scattered over the base matrix and a high density of Cr and C at the area where the carbides were detected. Therefore, the carbide was judged to be chromium carbides. It also can be suggested that surroundings of the carbide contained low Cr and C.

Table 4 showed the chemical compositions of specimens. The test surfaces were finished to 0.2 μm Ra of surface roughness to equalize the surface conditions before the erosion tests.

2.3 High temperature Vickers hardness test

The hardness test was conducted using AVK-HF type high temperature hardness tester. Specimen was measured to $6.5 \times 6.5 \times 5.0$ mm, and test surface was finished with polishing process. Test temperatures were room temperature and 1173 K. The temperature error range was assumed to be ±5 °C. Temperature increasing speed was 10 °C per minute. Hardness test was conducted after keeping on the certain test temperature for 5 min. Argon (or Ar) was blown into test chamber to control the oxidation of test surface. The diamond indenter was used at a load of 1 kgf with 10 s loading. The average values of 10 reading points were noted.

In addition, JCM-5000 scan Electronic microscopy (a product of JEOL) was used to characterize their wear surfaces, and the cross-section observation was investigated by VHX-100 series digital microscope (a product of Keyence).

3. Results and discussions

3.1 Difference of wear removal at room and high temperature

Erosion tests at room temperature (abbrev. R. T.) and 1173 K are carried out for SUS410. The testing conditions are shown in Table 3. Fig. 6 shows the erosion rate as a function of impact angle both at room temperature and 1173 K. Obviously, erosion rate at impact angle of 30° was 0.207×10^{-3} cm³/kg at room temperature, however, when the temperature was increased to 1173 K, the erosion rate reaches 7.90×10^{-3} cm³/kg at 30°. This indicated that the erosion rate increases approximately 40 times when testing temperature was increased from room temperature to 1173 K.

The surface morphologies of SUS410 at room temperature and 1173 K are determined using scan electron microscopy. Fig. 7 shows

151

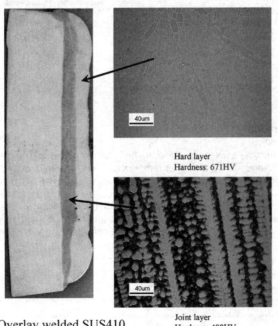
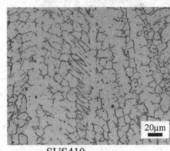
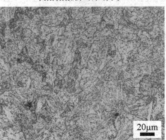

Overlay welded SUS410

Hard layer
Hardness: 671HV

Joint layer
Hardness: 488HV

SUS410
Hardness: 273HV

Forged SUS410
Hardness: 171HV

Fig. 4. Microstructure of specimens.

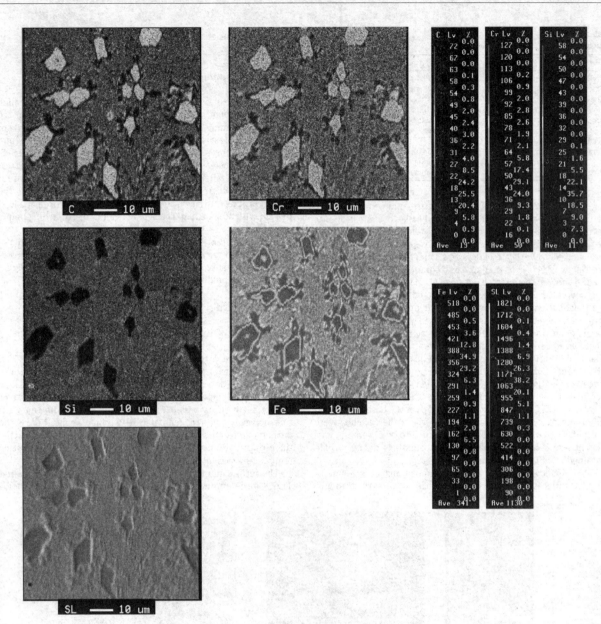

Fig. 5. EPMA surface analysis for overlay welded SUS410.

the surface morphologies of SUS410 after erosion test at impact angle of 30° both for room temperature and 1173 K. The figures illustrated that well-developed ripple patterns lying transverse to impact direction of erodent are observed on the test surface, where no evidence of ripples at room temperature, when test temperature reached 1173 K. That means surface structure was softening by high temperature to display ductile property. From the large magnified pictures in Fig. 7, it can be seen that cutting wear occurred at the eroded surfaces and this evidence became severe when temperature was increased. Accordingly, the target material showed maximum value at shallow angle of 30° just like ductile materials [1]. Therefore, in order to restrain the high temperature erosion, it was extremely important that controlling the wear removal at impact angle of 30°. Overlay welding method and forging process were considered to the optional cost effective way.

Table 4
Chemical compositions of specimens.

	C	Si	Mn	Ni	Cr	P	S
SUS410	0.13	0.28	0.38	0.27	11.64	0.027	0.017
Main components of overlay welded SUS410	Hard layer	Fe, Cr, C, Si					
	Joint layer	Fe, Mn, Ni					

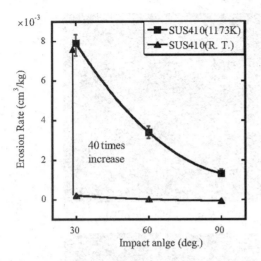

Fig. 6. Differences of ambient and high temperature erosion removal for SUS410.

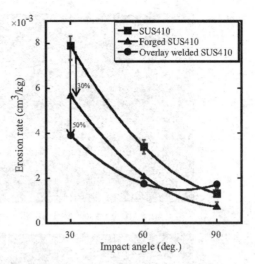

Fig. 8. Erosion rate as a function of impact angle at 1173 K for specimens.

3.2 Erosion rate and impact angle dependency

Fig. 8 shows the relationship between the erosion rate and the impact angle for forged SUS410, overlay welded SUS410 and SUS410 at 1173 K. The erosion rates reach the maximum values at the shallow angle of 30°, and then dramatically decrease with angle, typical erosion behaviour of the ductile material, for all specimens. It suggested that specimens were softened by high temperature to become ductile.

The figure also revealed that the erosion rate of forged SUS410 is lower than that of SUS410 at all of impact angles. Especially at 30°, erosion rate is reduced by approximately 30%. It suggested that forging process could control the erosion rate of base metal. The erosion rates of overlay welded SUS410 is twice that of the base metal of SUS410 both at impact angles of 30° and 60°. While the

153

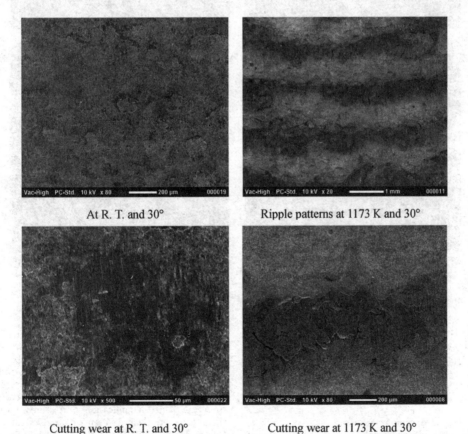

At R. T. and 30°

Ripple patterns at 1173 K and 30°

Cutting wear at R. T. and 30°

Cutting wear at 1173 K and 30°

Fig. 7. Surface morphology differences of SUS410 after erosion test both at room temperature (R. T.) and high temperature of 1173 K.

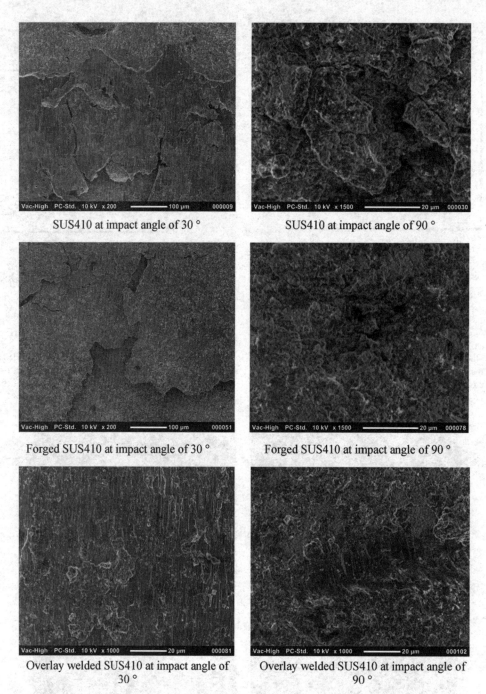

SUS410 at impact angle of 30 °

SUS410 at impact angle of 90 °

Forged SUS410 at impact angle of 30 °

Forged SUS410 at impact angle of 90 °

Overlay welded SUS410 at impact angle of 30 °

Overlay welded SUS410 at impact angle of 90 °

Fig. 9. Surface morphologies of samples after erosion test at 1173 K.

erosion rate of overlay welded sample at 90° is larger than that of it base metal, indicating that hard carbide structure of the surface of overlay welding affects the higher erosion rate at such a high impact angle.

3.3 Observation of surface morphologies

Surface morphologies of samples are characterized with SEM. Fig. 9 shows the eroded surfaces morphologies of SUS410, forged SUS410 and overlay welded SUS410 at 30° and 90° after ero-

sion test at high temperature of 1173 K. The figures revealed that impacted by alumina ball at high temperature of 1173 K, SUSU410 and forged SUS410 show clear ripple pattern at the impact angle of 30°. This evidence was illustrated in Fig. 7. All these suggested that there easily occurred plastic deformation due to high temperature softening of structure. Furthermore, from the large magnification of pictures revealed in Fig. 9, there have no clear differences observed at the eroded surface of 30° for SUS410 and forged SUS410. For impact angle of 90°, there have developed random indentations but no coher-

154

Table 5
Vickers hardness differences at room and high temperature of 1173 K.

	Surface hardness at room temperature	Surface hardness at 1173 K	Subsurface hardness (100–200 μm from eroded surface)
SUS410	273 HV	70 HV	245
Forged SUS410	171 HV	50 HV	280
Overlay welded SUS410	671 HV	184 HV	–

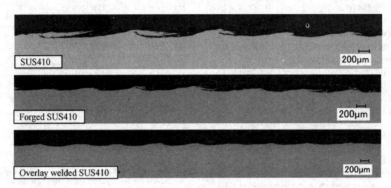

Fig. 10. Cross-section images of SUS410, forged SUS410 and overlay welded SUS410 at impact angle of 30°.

ent ripple pattern at the eroded surfaces of SUS410 and forged SUS410.

As displayed in Fig. 9, for overlay welded SUS410, a nearly plain eroded surface with small cutting traces was formed at the impact angle of 30°, and larger craters on the eroded area which were surrounded by relatively plain area, compare to other two materials, at the impact angle of 90°. This evidence indicated that the overlay welded hard layer may be contributed to restrain the plastic flow at high temperature.

3.4 Relationships between erosion rate and high temperature hardness

The results of hardness tests of SUS410, forged SUS410, and overlay welded SUS410 at room temperature and 1173 K are shown in Table 5. The table revealed that the Vickers hardness of SUS410 is dramatically decreased from 273 HV to 70 HV when room temperature rises to 1173 K. Similarly, hardness of forged SUS410 decreases to 50 HV from its initial hardness of 171 HV when temperature rises. That is to say, hardness at high temperature decreases approximately 70% from room temperature hardness for each material. This fact can explain the ductile property and high erosion rate at shallow angle of incidence. However, forged SUS410 shows lower erosion rate than SUS410 at high temperature, even if the hardness of forged SUS410 decreased more than SUS410. It suggested that erosion rate did not just depend on high temperature hardness at high temperature region; there may be other factors that could influence on high temperature erosion removal of materials. Accordingly, the characters of subsurface areas under the eroded surfaces of SUS410 and forged SUS410 were investigated. It was generally accepted that cold forging process generates the compressive residual stress at the surface and subsurface structure of material. There is a proportional relationship between the compressive residual stress and hardness [18]. As for the high temperature erosion, in this study, the test surfaces were soften and generated ripple patterns, assuming that the residual stress was relaxed by the comprehensive effects of the tensile and compressive stresses that were generated at the crests and troughs of the ripples. And these effects are some for SUS410 and forged SUS410. Therefore, the hardness of subsurface 100–200 μm away from the eroded surface was measured after erosion tests. Table 5 shows the results. It can be seen that hardness of subsurface of forged SUS410 was measured to be an average value of 280 HV, which was harder than that of SUS410 at the same area. This led to a suggestion that the compressive residual stress of forged SUS410 was still remained at the high temperature of 1173 K which contributing to the lower erosion rate of it.

The Vickers hardness of overlay welded material becomes 184 HV from its room temperature hardness of 671 HV when temperature reaches 1173 K. Although the hardness of the materials decreased dramatically when temperature increased, without expectation of the overlay welding material which show better erosion rate, the hardness at 1173 K is 3 times harder than that of SUS410. It suggested that high hardness carbides dispersed on the

Table 6
Wave length, wave depth and protrusion length of eroded surface of samples.

		Average value (μm)	Maximum value (μm)	Minimum value (μm)
SUS410	Wave length	1566	2113	1173
	Protrusion length	570	1048	64
	Wave depth	122	225	70
Forged SUS410	Wave length	1295	1636	970
	Protrusion length	263	496	44
	Wave depth	98	136	61
Overlay Welded SUS410	Wave length	939	1060	830
	Protrusion length	–	–	–
	Wave depth	56	71	43

surface of the overlay welded material contribute to the high temperature hardness, controlling of plastic flow and restraining the wear loss at the high temperature region.

3.5 *Cross-section observation*

Fig. 10 shows the cross-section pictures of SUS410, forged SUS410 and overlay welded SUS410 at impact angle of 30°. Table 6 shows the results of wavelength, wave depth and protrusion formed at the crests of ripples in samples.

It has been recognized by many researchers [1–5] that test surface of specimen was flown and extruded plastically in the direction of impact to form protruding portions at the crests of ripples. The protruding portions were grown by subsequent impact of particle and were broken off eventually. The larger the protruding portion, the easier to break off and the larger amount of erosion rate. Both Fig. 10 and Table 6 revealed that SUS410 has not only longer wavelength and deeper wave length but also longer protruding portions than that formed on the eroded surface of the forged SUS410 and the overlay welded SUS410. The cross-section of overlay welded SUS410 also revealed relatively plain surface and little protruding portions formed. These also suggested that compressive residual stress and higher hardness may likely to contribute to the controlling of plastic flow of surface.

4. Conclusions

This study aimed to the surface modification as a countermeasure to the erosion wear in the high temperature region, and high temperature erosion tests were performed on these surface modified materials and relative analysis also done. The result is shown as follows.

1. At the high temperature of 1173 K, the erosion at lower impact angle for the materials are increased dramatically because decrease of high temperature hardness make the material to be ductile and the cutting wear is remarkably appeared, especially at impact angle of 30°.
2. Erosion rate of forged 410 was controlled, suggesting that compressive residual stress may likely be remained in the subsurface structure that increases the hardness of under surface even if the surface hardness is softer than SUS410.

3. The larger wave length, wave depth and protruding portion make larger amounts of erosion rate, suggested that compressive residual stress and higher hardness may likely to result in the controlling of plastic flow of surface and erosion loss.

References

[1] I. Finnie, Erosion of surfaces by solid particles, Wear 3 (1960) 87–103.
[2] K. Shimizu, T. Noguchi, T. Kamada, H. Takasaki, Formation and progression of erosion surface in spheroidal graphite cast iron, Transactions of the American Foundrymen's Society 104 (1996), paper 96–61, 511–515.
[3] K. Shimizu, T. Noguchi, Erosion characteristics of ductile iron with various matrix structures, Wear 176 (1994) 255–260.
[4] X. Yaer, K. Shimizu, et al., Erosive wear characteristics of spheroidal carbides cast iron, Wear 264 (2008) 947–957.
[5] M.A. Uusitalo, P.M.J. Vuoristo, T.A. Mäntylä, Elevated temperature erosion–corrosion of thermal sprayed coatings in chlorine containing environments, Wear 252 (2002) 586–594.
[6] D.E. Alman, J.H. Tylczak, J.A. Hawk, J.H. Schneibel, An assessment of the erosion resistance of iron-aluminide cermets at room and elevated temperatures, Materials Science and Engineering A329–A331 (2002) 602–609.
[7] N. Hayashi, K. Hasezaki, S. Takaki, High-temperature erosion rates of Fe–Cr–C alloys produced by mechanical alloying and sintering process, Wear 242 (2000) 54–59.
[8] N. Hayashia, Y. Kagimoto, A. Notomia, Y. Takeda, K. Kato, Development of new testing method by centrifugal erosion tester at elevated temperature, Wear 258 (2005) 443–457.
[9] E. Vietzke, V. Philipps, Investigation of the high temperature erosion of nickel under 5 keV neon irradiation, Journal of Nuclear Materials 337–339 (2005) 1024–1028.
[10] D.W. Wheeler, R.J.K. Wood, Erosion of hard surface coatings for use in offshore gate valves, Wear 258 (2005) 526–536.
[11] M. Divakar, V.K. Agarwal, S.N. Singh, Effect of the material surface hardness on the erosion of AISI316, Wear 259 (2005) 110–117.
[12] S.B. Mishra, S. Prakash, K. Chandra, Studies on erosion behaviour of plasma sprayed coatings on a Ni-based superalloy, Wear 260 (2006) 422–432.
[13] B.-Q. Wang, A. Verstak, Elevated temperature erosion of HVOF Cr3C2/TiC–NiCrMo cermet coating, Wear 233–235 (1999) 342–351.
[14] B.Z. Janos, E. Lugscheider, P. Remer, Effect of thermal aging on the erosion resistance of air plasma sprayed zirconia thermal barrier coating, Surface and Coatings Technology 113 (1999) 278–285.
[15] S. Matthews, B. James, M. Hyland, High temperature erosion of Cr3C2-NiCr thermal spray coatings — the role of phase microstructure, Surface and Coatings Technology 203 (2009) 1144–1153.
[16] J. Barber, B.G. Mellor, R.J.K. Wood, The development of sub-surface damage during high energy solid particle erosion of a thermally sprayed WC-Co-Cr coating, Wear 259 (2005) 125–134.
[17] K. Shimizu, T. Naruse, Y. Xinba, K. Kimura, K. Minami, H. Matsumoto, Erosive wear properties of high V–Cr–Ni stainless spheroidal carbides cast iron at high temperature, Wear 267 (2009) 104–109.
[18] T. Shuji, Y. Yasuo, On the relation between residual stress and micro indentation hardness, Journal of the Japan Society for Testing Materials 11 (105) (1962) 386–391.

Contents lists available at ScienceDirect

Wear

journal homepage: www.elsevier.com/locate/wear

Erosive wear properties of high V–Cr–Ni stainless spheroidal carbides cast iron at high temperature

K. Shimizu[a,*], T. Naruse[b], Y. Xinba[a], K. Kimura[c], K. Minami[c], H. Matsumoto[d]

[a] Muroran Institute of Technology, Japan
[b] National Maritime Research Institute, Japan
[c] Nichias Corporation, Japan
[d] Sankyo Co. Ltd., Japan

ARTICLE INFO

Article history:
Received 1 September 2008
Received in revised form 21 December 2008
Accepted 21 December 2008

Keywords:
Erosive wear
Spheroidal carbide
High temperature
Hardness

ABSTRACT

To investigate the high temperature erosive wear properties of materials, a new testing machine was developed, and its performance was evaluated.

In this machine, temperature of impact particles, air, and test pieces were all controlled independently from room temperature to 900 °C, respectively. And the test pieces were set in a chamber to avoid occurring oxidation at high temperature. Heated particles were accelerated by hot air of same temperature level to impact on test piece that was heated up to same temperature level, so testing temperature condition was remained constant. The impact angles of test piece stage can be easily adjusted from 0° to 90°. Some experiments were performed using this testing machine. Erosive wear resistances of S50C steel, SK3 and high V–Cr–Ni stainless spheroidal carbides cast iron (SCI-VCrNi) were evaluated. Surface morphology of each material was characterized with optical microscope and micro-hardness tester.

Erosive wear resistance was related to both base matrix structure and material removal mechanism. The erosion rates of specimens increased with test temperature, among them, the erosion rate of SCI-VCrNi was the lowest. As a result, utilization of this erosion machine was confirmed to an effective way for the evaluation of high temperature erosion.

157

1. Introduction

Erosive wear by solid particles impact has been a serious problem for many industrial components, mainly at bended sections of pipe, valve, turbine blade, the fan in a pneumatic conveying system, even the blade of helicopter, etc. [1–3].

Instances of high temperature erosive wear occur in production of the inorganic fibrous insulator in plant. The inorganic fibrous insulator has been widely used in the construction materials, internal insulation of industrial furnace, power generation industry and space industry because of its light-weight, high insulation ability and cost-effective manner. However, in the manufacturing processes, the melting slag (wool) impacts on highly rotating rotor and stretches out the melting slag by centrifugal force. During the procedure, the rotors are eroded by impact of melted slag. To improve the yield rate and production quality of inorganic fibrous insulator, increasing of centrifugal force is an effective way. When increasing the number of revolution, however, the life span of rotor decreases to 1/10–1/2 of the former one, which makes increasing of rotor exchange frequency and costs. Therefore,

developing of the rotor materials are an important problem to be solved.

Accordingly, mechanism of materials in high temperature region has to be solved and demands for development of excellent wear and thermal resistant materials are on the increase. So, in order to predict and evaluate the erosion of rotors, it is not only necessary to develop the new method of particle erosion testing under high temperature conditions, but to manufacturing a new material for high temperature.

On one hand, there are three types of setups for a particle erosion test at high temperature used by research institutes, such as rotating arm type, blast type and wind tunnel type. Most institutes have used blast type apparatuses, especially blowing particles by a gas jet [6].

On the other hand, many of researchers [13–16] mainly aimed at the erosive wear of composite material, superalloy, hard surface coatings, high chromium white cast iron, and so on; but they are seem to not cost effective materials. Among them, high chromium white cast iron has been defined as a cast iron with good wear-resistance in abrasive condition, however, the presence of dendrite or flaky shaped carbides embedded in matrix drew them occur stress concentration easily to limit their wear resistance.

In the base of mentioned above, we developed the cast irons with spheroidal carbides embedded in matrix. The cast irons were obtained in the pouring temperature of 1700–1800 °C with adding

* Corresponding author. Tel.: +81143465651; fax: +81143465651.
E-mail address: shimizu@mmm.muroran-it.ac.jp (K. Shimizu).

0043-1648/$ – see front matter © 2009 Elsevier B.V. All rights reserved.
doi:10.1016/j.wear.2008.12.086

approximately 10% vanadium to crystallize vanadium spheroidal carbides in the structure. The spheroidal vanadium carbides (VC) uniformly dispersed in austenite matrix and the room temperature erosion property of them had been made clear [1]. The erosion property at high temperature of these materials, however, are not generally appreciated or understood.

In the present research, the structure of an originally designed high temperature experimental setup and procedure of high temperature erosion tests are described. The erosion properties of high V–Cr–Ni stainless spheroidal carbides cast iron with comparison of SK3 and S50C at high temperature region are studied.

2. Experimental procedure

2.1 *Structure of new high temperature erosion setup*

A high temperature erosion setup was built to test the erosive wear of target materials in the present investigation. Fig. 1 shows the picture of the new high temperature erosion setup, and the schematic diagram of the new setup is shown in Fig. 2. Particles are supplied from the particle chamber 1 into the chamber 2 where entrained by hot air to fly out from the nozzle. The setup has following characteristics.

Fig. 1. Picture of blast type high temperature erosion setup.

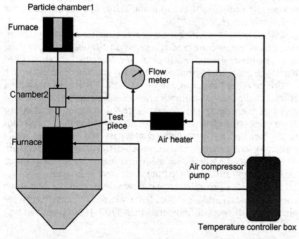

Fig. 2. Schematic diagram of the high temperature erosion setup.

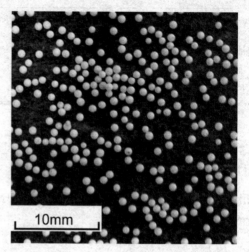

Fig. 3. Appearance of erodent particle.

(1) Test piece is set in a chamber to avoid occurring oxidation at high temperature. Heated particles are accelerated by hot air of same temperature level to impact on test piece that is also heated up to the same temperature level, so testing temperature condition remains constant.
(2) Since temperature of test piece, erodent particles and air are controlled independently; data of erosion under different temperature conditions can be obtained.
(3) Impact angle of test piece stage can easily be adjusted from 0° to 90°. So erosion data at all of angle can be obtained.

The impact particles are changed after each erosion test. Specimens with dimensions $50\,\text{mm} \times 50\,\text{mm} \times 10\,\text{mm}$ are used for the high temperature erosion test. The specimens are mounted into the test stage directly below the nozzle with a vertical distance of 50 mm from the end of the nozzle to the test surface into the erosion test machine by changing their impingement angles 30°, 60°, and 90°, respectively. The examined air max speed is 226 m/s, and the particle feed rate is approximately 26 g/s.

2.2 *Erodent particles*

Spherical shaped alumina ball with average diameter 1 mm, 9 Mohs hardness are used as the erodent material. Erosion test duration is 300 s and the amount of erodent is 8 kg/test. During the high temperature erosion test, the erodent is heated to 1173 K, so other erodent such as steel grits or silica sand, which were generally used as erodent for erosion tests [3–7], would be deformed or fractured by heat and blasting. The selected alumina ball has excellent wear and thermal resistance. The particles appearance of the erodent material is shown in Fig. 3. Some characteristics of the erodent material are listed in Table 1.

Table 1
Specification of erodent particle.

Erodent particle	Alumina ball
Specific gravity	3.6 or greater
Mohs hardness	9
Components	Alumina purity 92% or greater
Average size (mm∅)	1

Table 2
Summary of erosion test conditions.

Test variable	Test 1	Test 2
Erodent particle temperature, T (K)	873	1073
Test piece temperature, T (K)	1173	1173
Entrained air temperature, T (K)	1073	1073
Hot air velocity, V (m/s)	226	226
Impact angle, α (°)	90	90
Test duration, t (s)	300	300
Erodent particle loading, L (kg)	8	8
Target materials	SS400	SS400
Erodent particle	Alumina ball (1 mm Ø)	Alumina ball (1 mm Ø)

2.3 Assessment method

Specimens are weighed with an electronic scale both before and after the test. The removed material volume is calculated as unit amounts of particles impact on specimen surface, which is defined as erosion rate, to compare those materials having different densities, as proposed by Finnie [2]. The formula of erosion rate is as follow:

$$\text{Volumetric removal per second (cm}^3/\text{s)} = \frac{\text{Mass removal per second (g/s)}}{\text{Average density (g/cm}^3)}$$

$$\text{Erosion rate (cm}^3/\text{kg)} = \frac{\text{Volumetric removal per second (cm}^3/\text{s)}}{\text{Mass amounts of impact particles per second (kg/s)}}$$

The reason why is the volumetric removal divided by mass amounts of impact particles is that the mass removal per second should depend on the total amounts of particles that impact on target material in that period. It is considered that the more the amounts of particles impact on surface of material the more the material removal increased. So it is more reasonable to determine the value at amount of impact particles per second as erosion rate. After erosion test, the specimen is again weighed to an accuracy of 0.01 mg, to determine the mass loss. The macrostructure of eroded surfaces are also observed [8–12].

2.4 Effect of erodent particles temperature on erosion removal

A large number of researches were found in the literature of temperature effect on materials [3–7]. Many researchers have paid more attention on high temperature testing conditions, especially on temperature of test piece. In this study, we investigate the temperature of erodent particles on a same test piece.

The relationship between the erosion removal and temperature of erodent particles for SS400 is tested. The testing conditions are shown in Table 2. The erosion results are shown in Fig. 4. When the erodent particles temperature is 873 K, erosion rate of SS400 is 41.148×10^{-3} cm^3/kg, however, when the erodent particles temperature is increased to 1073 K with no change of other testing conditions, the erosion rate reaches 61.854×10^{-3} cm^3/kg. Erosion rate increases 1.5 times with increasing temperature of erodent particles from 873 to 1073 K. It is supposed that the influence of high temperature softening at 1073 K is larger than at 873 K condition, so the erosion rate is more strongly increased.

Therefore, setting up the erosion test conditions is extremely important for high temperature erosion test. It is needed to keep the whole temperature at same level.

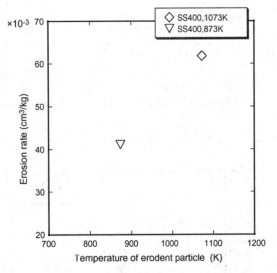

Fig. 4. Erosion rate vs. erodent particle temperature in SS400.

3. Results and discussion

3.1 Procedure to setup erosion test conditions

From above-mentioned, it is important to maintain the erosion test conditions at a same level. For this reason, the measurement of particle velocity, particle impingement angle, especially, temperature of erodent particles, hot air and a test piece are necessary. In this investigation, the high temperature erosion setup is possible to realize keeping all temperature conditions at a same level. A summary of test conditions is shown in Table 3. Erosion tests were conducted in a high temperature blast typed erosion tester. Hot air was utilized as the carrier gas for the particles. Particle velocity was controlled by setting pressure drop across the nozzle using a metering system that was connected to an air compressor.

3.2 Preliminary tests for reproducibility of high temperature erosion setup

To confirm the reproducibility of the original high temperature erosion setup, we use a test piece of SS400 to do high temperature erosion tests. The testing conditions are shown in Table 3. The tests in same condition are conducted for three times to confirm the reproducibility of erosion results. Fig. 5 shows their erosion rates. Form the values obtained from three tests, it can be confirmed that the erosion rates obtained by the high temperature erosion apparatus are excellently reproducible. Utilization of this erosion setup is confirmed to an effective way for the evaluation of high temperature erosion.

Table 3
High temperature erosion test conditions.

Test variable	Test conditions
Erodent particle temperature, T (K)	1173
Test piece temperature, T (K)	1173
Entrained air temperature, T (K)	1073
Hot air velocity, V (m/s)	226
Impact angle, a (°)	90
Test duration, t (s)	300
Erodent particle loading, L (kg)	8
Target materials	SS400, SK3, S50C, SCI-VCrNi
Erodent particle	Alumina ball (1 mm Ø)

159

Table 4
Chemical compositions and mechanical properties of alloys.

	C	Si	Mn	Ni	Cr	Nb	V	Others
S50C	0.47–53	0.15–0.35	0.60–0.90	Maximum 0.200	Maximum 0.200	Maximum 0.35	–	P, S, Cu
SK3	1.0–1.1	0.1–0.35	0.1–0.5	–	–	–	–	P, S, Cu
SCI-VCrNi	3	1	0.7	8	18		10	–

	σ_B	ϕ	HV
S50C	705–900	15	250
SK3	850 or greater	–	560
SCI-VCrNi	650–750	5	390

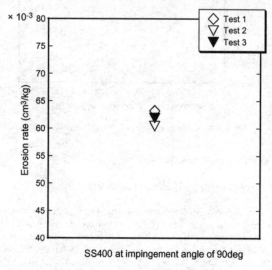

Fig. 5. Erosion rate reproducibility of SS400.

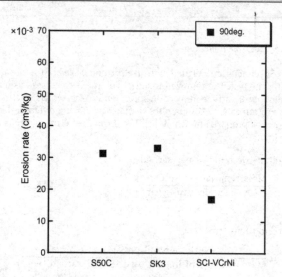

Fig. 7. Erosion rate vs. impact angle of materials at 1173 K.

3.3 Erosion test results

In order to examine high temperature erosion removal of test pieces, erosion tests on three kinds of materials were carried out using the new high temperature blast typed erosion setup under test condition showed in Table 3. Alloys used for high temperature erosion tests were S50C steel, SK3 and high V–Cr–Ni stainless spheroidal carbides cast iron (SCI-VCrNi), respectively. Table 4 shows chemical compositions and mechanical properties of the materials. Fig. 6 shows their microstructure. Fig. 7 shows the erosion rate results of specimens at 1173 K for impingement angle of 90°. It can be seen that S50C and SK3 were most eroded, and among them, the erosion rates of SCI-VCrNi were the lowest. Table 5 summarizes microstructural features, hardness, and erosion rate data for all materials. As a result, it can be concluded that SCI-VCrNi is an effective wear and thermal resistant casting materials.

Fig. 6. Microstructure of specimens ((A) S50C, (B) SK3, (C) SCI-VCrNi).

Table 5
Summary of materials some features and erosion rate data.

Materials	Density (g/cm^3)	Vickers hardness	Temperature (K)	Test duration (s)	Erosion rate ($\times 10^3$ cm^3/kg)
S50C	7.81	250	1173	300	31.38
SK3	7.69	560	1173	300	33.05
SCI-VCrNi	7.04	390	1173	300	16.97

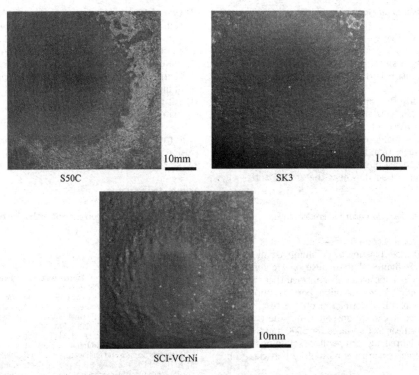

S50C

SK3

SCI-VCrNi

Fig. 8. Eroded surface observation of specimens after 300 s of blasting for 90°.

161

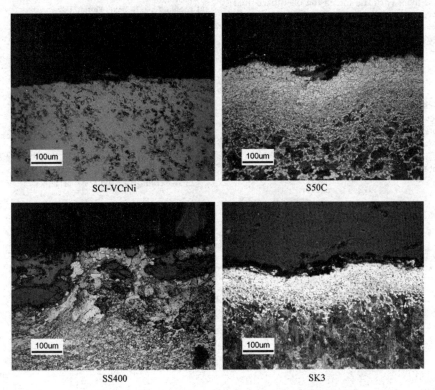

SCI-VCrNi

S50C

SS400

SK3

Fig. 9. Vertical section near eroded surface of specimens after 300 s of blasting for 90°.

3.4 *Test surface morphology and erosion mechanism*

To make clear the erosive wear behaviours of selected specimens, Surface morphology of each material is characterized with optical microscope. Fig. 8 shows the test surfaces of three specimens eroded by alumina ball at a 90°-impact angle after 300 s blasting.

It can be seen clearly that all the test pieces are eroded by impacting of erodent particles, among them SCI-VCrNi's eroded area is the smallest. SCI-VCrNi is thought to be a good wear and thermal resistant casting material. However, the reason why SCI-VCrNi resists erosion more than the other alloys is not clear just through surface morphology observation. Therefore, we go into observation on the vertical section near the eroded surface by microscope.

3.5 *Observation of vertical section near the eroded surface*

Fig. 9 shows the vertical section near test surface of SCI-VCrNi, S50C, SK3 and SS400 after blasting with alumina ball at 90°. We observe from these figures that because only compressive force stress on the material surface, it is confirmed that there are formed the compressed piliform protruding portions and indentations resulting from plastic deformation on the test surfaces of specimens. Moreover, it can be observed that there are the structure compressions near test surfaces for S50C and SK3. It is suggested that there occurred high temperature softening of structure and then the soften structure is blasted by particles to be compressed.

However, the structure compression near test surface is not occurred for SCI-VCrNi. It is supposed that when particles impact on the surface the austenite in matrix in SCI-VCrNi change to martensite by strain induced transformation, which results in increasing of hardness although high temperature softening of structure occurs in the meantime. And then, with the increasing of impact time, formed martensite would be deformed and stripped from surface, simultaneously, the structure under surface again change to martensite repeatedly to form new hard layer and this layer maintain certain thick from the eroded surface.

4. Conclusions

The present work is intended to establish a new high temperature erosion test and evaluate erosion rates of some materials under high temperature condition. From results of erosion tests by the new testing method the following conclusions have been obtained.

(1) The erosion rate obtained by the high temperature erosion setup is reproducible.

(2) One of the characteristics of the new high temperature erosion setup is that temperature of erodent particle, test piece and entrained air were controlled separately and erosion test condition can be controlled to same level during test. So the obtained data from this setup is more accurate.

(3) High temperature erosion is extremely sensitive to temperature at high temperature level.

(4) SCI-VCrNi is an excellent wear and thermal resistant casting material.

In the next step experiment, effect of impingement angle on erosion rate, particle impingement velocity dependency will be investigated more closely.

Acknowledgements

This research has been supported by Sankyo Co. Ltd. and Nichias Corporation Japan.

References

[1] Y. Xinba, K. Shimizu, et al., Erosive wear characteristics of spheroidal carbides cast iron, Wear 264 (2008) 947–957.
[2] I. Finnie, Erosion of surfaces by solid particles, Wear 3 (1960) 87–103.
[3] M.A. Uusitalo, P.M.J. Vuoristo, T.A. Mäntylä, Elevated temperature erosion–corrosion of thermal sprayed coatings in chlorine containing environments, Wear 252 (2002) 586–594.
[4] D.E. Alman, J.H. Tylczak, J.A. Hawk, J.H. Schneibel, An assessment of the erosion resistance of iron-aluminide cermets at room and elevated temperatures, Materials Science and Engineering A329-331 (2002) 602–609.
[5] N. Hayashi, K. Hasezaki, S. Takaki, High-temperature erosion rates of Fe–Cr–C alloys produced by mechanical alloying and sintering process, Wear 242 (2000) 54–59.
[6] N. Hayashia, Y. Kagimoto, A. Notomia, Y. Takeda, K. Kato, Development of new testing method by centrifugal erosion tester at elevated temperature, Wear 258 (2005) 443–457.
[7] E. Vietzke, V. Philipps, Investigation of the high temperature erosion of nickel under 5 keV neon irradiation, Journal of Nuclear Materials 337–339 (2005) 1024–1028.
[8] K. Shimizu, T. Noguchi, Erosion characteristics of ductile iron with various matrix structures, Wear 176 (1994) 255–260.
[9] K. Shimizu, T. Noguchi, Fundamental study on erosive wear of austempered ductile iron, in: Proceeding of the 3rd East Asian International Foundry Symposium, Pusan, Korea, 1–3 July, 1992.
[10] K. Shimizu, T. Noguchi, S. Doi, Basic study on the erosive wear of austempered ductile iron, Trans. AFS 101 (1993) 225–229.
[11] K. Shimizu, T. Noguchi, T. Kamada, H. Takasaki, Progress of erosive wear in spheroidal graphite cast iron, Wear 198 (1996) 150–155.
[12] K. Shimizu, T. Noguchi, T. Kamada, S. Doi, Basic study of erosion of ductile iron, Adv. Mater. Res. 4–5 (1997) 239–244.
[13] I. Hussainova, Microstructure and erosive wear in ceramic-based composites, Wear 258 (2005) 357–365.
[14] D.W. Wheeler, R.J.K. Wood, Erosion of hard surface coatings for use in offshore gate valves, Wear 258 (2005) 526–536.
[15] M. Divakar, V.K. Agarwal, S.N. Singh, Effect of the material surface hardness on the erosion of AISI316, Wear 259 (2005) 110–117.
[16] S.B. Mishra, S. Prakash, K. Chandra, Studies on erosion behaviour of plasma sprayed coatings on a Ni-based superalloy, Wear 260 (2006) 422–432.

162

研究論文

高マンガン球状炭化物鋳鉄のサンドエロージョン摩耗特性

新巴雅尔[*]　清水一道[*]　桃野　正[*]
松元秀人[**]　橘堂　忠[***]

Research Article
J. JFS, Vol. 78, No. 10 (2006) pp. 510～515

Erosive Wear Properties of High Manganese Cast Iron with Spheroidal Carbides

Xinbayaer[*], Kazumichi Shimizu[*], Tadashi Momono[*],
Hideto Matsumoto[**] and Tadashi Kitsudou[***]

Surface damage caused by the impact of dispersed particles in gas or liquid flow is called "erosion". Erosion is a serious problem as it occurs in piping systems due to gas-solid flow. This report discusses the evaluation of high manganese spheroidal carbide cast iron which is expected to have high resistance against erosive wear.

Erosive wear test was performed on flaky graphite cast iron (FC200), spheroidal graphite cast iron (FCD400), white cast iron (WCI), and high manganese cast iron with spheroidal carbides (SCI-VMn) using a shot blast machine. Erosion damage was measured by the removed material volume, defined as erosion rate at impact angle 30°, 60°, and 90°. By way of the mechanism of erosive wear, not only was the surface metal flow in vertical sections observed, but the effect of impact angles, and differences in wear features of specimens were also discussed.

Results of experiment showed that, the erosion rate for SCI-VMn is about 1/8 of that for FC200, and about 1/6 of that for FCD400. Although the initial hardness of WCI was highest among all specimens, its erosion rate was larger than SCI-VMn. Work hardening effect was seen in all specimens, especially, the surface hardness of SCI-VMn increased from the initial hardness HV530 to HV804 after 3600 sec. of blasting. The reason for this maybe that the austenite in the structure undergoes strain induced martensitic transformation which results in surface hardening and lower erosion rate. It was shown that SCI-VMn has excellent erosion resistance and it is expected to be available for wide applications as a wear-resistant material.

Keywords : wear, erosion, spheroidal carbides cast iron, VC

163

1. 緒　言

固体粒子によって生じる摩耗は，その摩耗形態の違いによって幾つかに分類されている．その中で，固体粒子が材料に衝突することにより，その材料の表面が損傷，除去される現象をエロージョンと呼んでいる[1]．エロージョン現象は，主に輸送系でのパイプベンド部やバルブ，タービンブレード，ヘリコプターのブレード及びファンなどにおいて大きな問題となっている[2~4]．現状では定期的なメンテナンス時に配管を交換したり，減肉部分に肉盛り溶接を施すことにより事故を回避しているが，未然に事故を防ぐためにも耐摩耗材料の開発及び余寿命の推定は解決すべき重要な課題である．

これまで，清水らの研究により軟鋼，球状黒鉛鋳鉄，ステンレス鋼及び高クロム鋳鉄などのエロージョン摩耗特性を明らかにしたが[2~9]，長寿命化を目的とした優れたエロージョン摩耗特性を示す材料は得られていない．そこで，本研究では，特に耐摩耗材料として期待される高マンガン球状炭化物鋳鉄に着目して固気二相流において生じるサンドエロージョンによる摩耗特性評価を行った．高マンガン球状炭化物鋳鉄とは，バナジウムを約10 mass%添加することで組織内に高硬度の球状のバナジウム炭化物（VC）が晶出した鋳鉄である．ねずみ鋳鉄の場合，組織が片状黒鉛とパーライトであるため，黒鉛部分が摩耗の起点となり，損傷へと進行するため，耐摩耗材としては，良好な結果は得られない．一方，バナジウムはクロムと同様に炭化物を安

平成 18 年 4 月 12 日　原稿受理
　　* 室蘭工業大学材料物性工学科　　Department of Material Science and Engineering, Muroran Institute of Technology
　 ** （株）三共合金鋳造所　　Sankyo Co. Ltd.
　*** 大阪府立産業技術総合研究所　　Technology Research Institute of Osaka Prefecture

定化する働きがあり，VC を球状に晶出させることにより，粒子衝突による応力を分散させ，材料の耐摩耗性が向上すると考えられる.

エロージョン摩耗の支配因子としては，使用材料の硬さと組織，衝突粒子の硬さ，粒径及び形状，さらに粒子の衝突速度及び衝突角度などの摩耗条件があげられ，材料の摩耗量，摩耗特性に大きな影響を及ぼすことが知られている[9]. 本研究では高マンガン球状炭化物鋳鉄のエロージョン摩耗に及ぼす衝突角度と衝突粒子2種類の影響を調べ，高マンガン球状炭化物鋳鉄の組織と硬さとエロージョン摩耗の関係を3種類の鋳造材と比較して検討した.

2. 実験材料及び実験方法

エロージョン摩耗試験の供試材として，片状黒鉛鋳鉄（FC200），球状黒鉛鋳鉄（FCD400），高クロム鋳鉄（WCI）及び基地中に球状炭化物を有する高マンガン球状炭化物鋳鉄（Spheroidal Carbides cast Iron : SCI-VMn）を用いた. それぞれの化学成分を **Table 1** に，供試材の組織を **Fig. 1** に示す.

エロージョン摩耗試験には，吸引式ブラストマシンを使用した. 試験片の寸法は 50×50×10 mm とし，衝突粒子には，スチールグリッド，球形，平均粒径 660 μm，硬さ 420 HV 及び硅砂，不定形，平均粒径 408 μm，硬さ 1030 HV を使用した. 衝突粒子はそれ自体も粉砕し，粒径が変化するため，試験毎に交換した. ノズルからの衝突速度は約 100 m/s，衝突角度は 30°，60°，及び 90° とし，噴出量はスチールグリッドで約 40 g/s，硅砂で約 4 g/s，試験時間は 3600 s. とした. 衝突粒子の総循環質量は 2 kg の条件で試験を行った. 摩耗質量は感度 0.01 mg の電子天秤により，試験前後の重量差から求めた. 本研究では密度の異なる材料を使用したため，その耐エロージョン摩耗性を比較する場合，材料の摩耗損傷は重量減少量で比較するのではなく，体積減少量で比較する方が適当である. そこで，摩耗を受ける材料の平均密度を用いて，損傷速度（Erosion rate, cm³/kg）で摩耗量を算出した[6]. 試験終了後，摩耗面の観察及び表面近傍の断面組織を観察した.

3. 実験結果

実験1では，粉粒体に球状のスチールグリッド，供試材に FC200，FCD400 及び SCI-VMn を用いてエロージョン摩耗実験を行った. このスチールグリッドを用いた理由は，形状が単純同一で，直径及び密度が一定であること，また，これまでの実験結果と比較するため，実験2における結果との比較が容易であることである. 各試料の衝突時間 3600 s. における衝突角度と損傷速度の関係を **Fig. 2** に示す. スチールグリッドで試験をした場合，SCI-VMn については，ほとんど摩耗せず，良好な耐エロージョン摩耗特性を示した. FC200 及び FCD400 においてエロージョン摩耗を示した. 両試料においては，30° の低角度側で摩耗量が少なく，角度が増すにつれ，損傷速度も増加してい

Table 1 Chemical compositions of various specimens. (Mass%)

Specimen	C	Si	Mn	Ni	Cr	V
FC200	3.24	1.79	0.26	0.06	—	—
FCD400	3.74	2.16	0.30	0.02	—	—
WCI	2.80	0.70	1.0	—	24.5	—
SCI-VMn	2.92	0.57	13.0	0.95	—	12.8

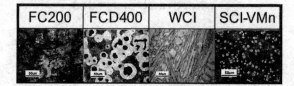

Fig. 1 Microstructure of specimens.

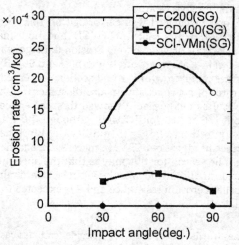

Fig. 2 Erosion rate vs. Impact Angle in FC200, FCD400 and SCI-VMn by steel grits.

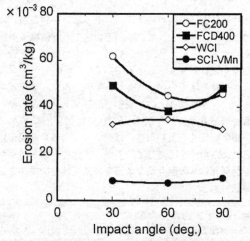

Fig. 3 Erosion rate vs. Impact angle in FC200, FCD400 and SCI-VMn by sand particles.

き，60°で最大の損傷速度を取る，その後は，角度が増す
につれ減少していき，90°では損傷速度は小さくなってい
る．

　スチールグリッドより高硬度で，不定形である硅砂を使
用し，先ほどと同様の供試材でエロージョン摩耗試験を行っ
た結果をFig.3に示す．高硬度かつ不定形の硅砂を使用
すると，FC200，FCD400及びWCIにおいて激しい摩耗
が起こり，SCI-VMnにおいても，摩耗が起っていたが，
明瞭なピークは現れなかった．このように，衝突する粒子
が変わると，摩耗量は著しく変化するだけではなく，材料
の衝突角度依存性も変化すると判断できる．これは清水ら
の報告[1]及びFinnieらの報告[10]と一致している．SCI-
VMnはFC200の損傷速度の1/8，FCD400の損傷速度
の1/6であり，耐摩耗材料として広く使用されている
WCIの1/4であり，衝突角度依存性もほとんど見られな
い．

4．摩耗メカニズム解析

4.1　表面マクロ観察

　各試料の摩耗機構を解明するために，実験後のエロージョ
ン摩耗面のマクロ観察を行った．Fig.4に衝突粒子にス

チールグリッド，Fig.5に硅砂を用いた場合のエロージョ
ン表面のマクロ写真を示す．以前の清水らの研究及び他の
研究者らの結果[11～17]によると，衝突粒子にスチールグリッ
ドを用いて実験した場合，S45Cにおいては，衝突粒子の
衝突方向と垂直に明瞭な縞模様が観察された．しかしなが
ら，本研究では，FC200及びFCD400において，S45C
のような明瞭な縞模様が観察されたが，SCI-VMnのよう
に衝突により硬度が著しく向上する材料では，衝突粒子の
衝突による痕がごくわずかに観察されるだけであって，
S45Cのような縞模様は現れなかった．衝突粒子に硅砂を
用いた場合のエロージョン摩耗表面のマクロ写真では，衝
突粒子にスチールグリッドを用いた場合のS45C[1]におい
て見られた縞模様が全ての試料のどの衝突角度においても
縞模様は現れない．一方SCI-VMnでは，衝突角度が増す
につれ，表面の摩耗面積が小さくなり，耐エロージョン摩
耗性に優れることが理解できる．

4.2　エロージョン断面ミクロ観察

　FCD400及びSCI-VMnが両方とも球状の組織を有する
ため，応力集中を分散させる効果が同じである，また
FCD400が組織内に球状黒鉛を含み，また黒鉛の周りに
パーライトを有するので，この球状黒鉛粒の硬さが低く，

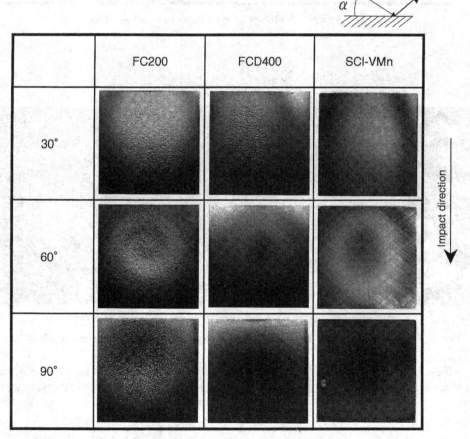

Fig. 4　Eroded surface in FC200, FCD400 and SCI-VMn by steel grits.

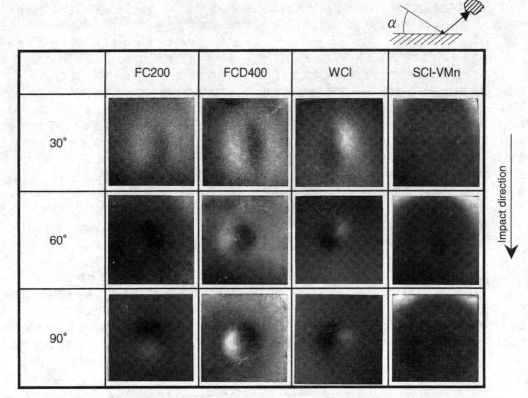

Fig. 5 Eroded surface of specimens by sand particles.

166

30°

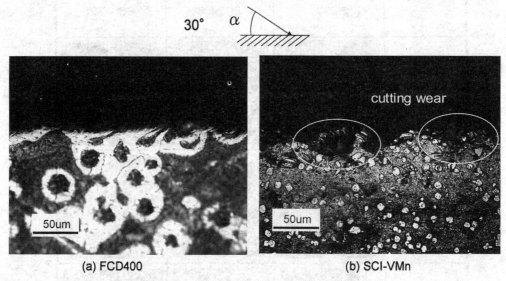

(a) FCD400　　　(b) SCI-VMn

Fig. 6 Cross section of FCD400 and SCI-VMn after erosion test at impact angles of 30 deg.

衝突によって塑性変形し，脱落してしまう，一方，SCI-
VMn の球状炭化物が非常に硬いので，応力集中を分散さ
せる効果がある上，塑性変形を押さえると考えられる．そ
こで，FCD400 を比較材料として SCI-VMn の断面ミクロ
観察をした．

Fig. 6，Fig. 7，及び Fig. 8 に衝突速度 100 m/s で衝
突粒子に硅砂を用いた場合の FCD400 及び SCI-VMn の
30°，60°及び 90°における断面写真を示す．まず，
FCD400 では，すべての角度において，表面近傍の黒鉛
が，変形，扁平化しており，塑性変形の跡が明瞭である．

学术论文

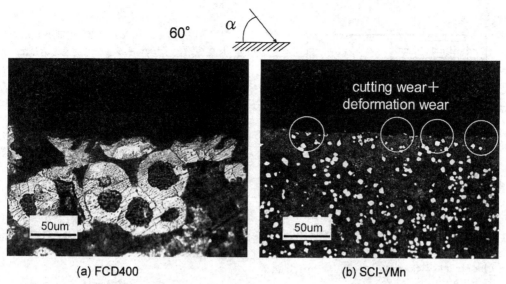

(a) FCD400 　　　　　　　　(b) SCI-VMn

Fig. 7　Cross section of FCD400 and SCI-VMn after erosion test at impact angles of 60 deg.

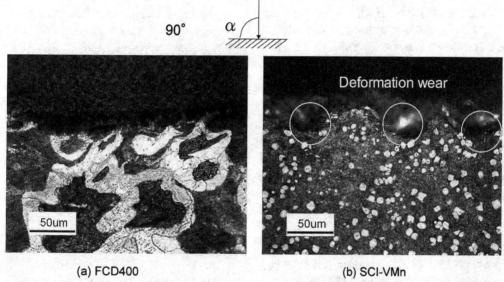

(a) FCD400 　　　　　　　　(b) SCI-VMn

Fig. 8　Cross section of FCD400 and SCI-VMn after erosion test at impact angles of 90 deg.

167

これに対して，SCI-VMnにおいては角度ごとに摩耗機構が異なることが理解できる．つまり球状炭化物が粒子の衝突による塑性変形に対して大きな抵抗力として働いていると考える．30°の低角度では材料表面に切削摩耗が生じ，小さな突起を形成している．球状炭化物は球状のまま形を留めており，周りのマトリックスだけが削れている．また，摩耗表面に残された球状炭化物鋳鉄が衝突方向に向かって塑性流動していない．これらのことから，球状のVCは非常に硬い上，マトリックスとの密着性も非常に良いことが理解できる．90°の場合では，材料表面には圧縮のみの力がかかるため，球状炭化物がマトリックスと同時に掘り起こされ，また，衝突粒子の衝突により，小さな突起部及び圧痕が形成される．さらに，球状炭化物が脱離する場合と割れが生じる場合の両方が確認されたが，球状炭化物に塑性変形は生じていない．マトリックスの硬化とともに，摩耗の進展を抑えられていた．また，60°の衝突の場合には，材料表面に30°のような切削摩耗と90°のような脱落や割れによる摩耗現象が同時に形成されることが観察された．

4.3　摩耗表面の硬さ測定

エロージョン摩耗試験では，材料表面に衝突粒子が衝突することにより，加工硬化されると考えられる．このことから，ビッカース硬さ試験により，各試料の試験前と試験後の硬さを測定した．

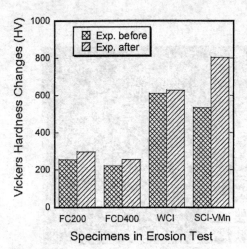

Fig. 9　Change of Vickers hardness for specimens after erosion test.

Fig.9 にその結果を示す．Fig.9 より，実験後，全ての試料において加工硬化を示した．最も損傷速度の小さかった SCI-VMn は，試験後の硬さ HV804 と試験前の硬さ HV530 と比べ，約 52% の硬さの上昇を示した．また，WCI の初期硬さは最も高い値を示したが，試験後はほとんど硬さが上昇していない．SCI-VMn の硬さが著しく上昇した理由は，マトリックスのオーステナイトがエロージョン摩耗進行時の衝突粒子により加工誘起変態してマルテンサイトに変わったためと考えられる．このように SCI-VMn の硬さが加工硬化で著しく上昇したため，SCI-VMn の耐摩耗性が向上したと考察できる．このことから，エロージョン摩耗は材料の硬さに依存しているが，加工硬化による硬さの上昇にも大きく依存していることが分かる．スチールグリッドを用いた場合で，摩耗量がわずかだった理由は衝突粒子の衝突によって SCI-VMn の表面マトリックスが加工誘起変態し，硬くなり，また，次の粒子の衝突によって破壊，または脱落されにくくなったと考えられる．硅砂の場合は，その加工誘起変態した部分が不定形の硅砂によってさらに破壊，脱落してしまい，摩耗が進むが，その部分の下部では再びマトリックスがマルテンサイト化によって，摩耗が抑えられたと理解できる．

エロージョン摩耗の損傷速度は衝突粒子の硬さや形状により異なるが，耐エロージョン摩耗性を向上させるには，硬質炭化物がマトリックス中に均一に分散し，マトリックスが加工誘起変態して，硬さが著しく増加するような材料を開発することが必要とされる．

5．結　言

本研究では，衝突粒子にスチールグリッド及び硅砂を用いて，FC200，FCD400，WCI 及び SCI-VMn の四種類の材料のエロージョン摩耗実験を行い，それら材料の耐エロージョン摩耗特性を評価した．その結果を以下に示す．
(1) 高マンガン球状炭化物鋳鉄 (SCI-VMn) は，最も摩耗

量が多かった FC200 に比べ損傷速度が 1/8 と小さく，非常に良好な耐エロージョン摩耗性を示した．その理由はマトリックスがオーステナイトであるため，エロージョン摩耗進行時の衝突粒子により，マトリックスがマルテンサイトに加工誘起変態したことで硬さが著しく上昇し，耐摩耗性が向上したと考察できる．
(2) エロージョン摩耗は，初期硬さよりも加工硬化による硬さの増加が重要である．加工誘起変態し，著しく硬さが増加する材料ほど，優れた耐エロージョン摩耗性を示す．

最後に，この研究に携わった室蘭工業大学大学院田中真人君，坂本卓君ならびに卒業生の立浪宏大君に感謝の意を表する．

参考文献

[1] Kazumichi Shimizu and Toru Noguchi : IMONO **66** (1994) 7
[2] Kazumichi Shimizu and Toru Noguchi : Trans. Jpn. Soc. Mech. Eng. 65-632A (1999) 940
[3] K. Shimizu, T. Noguchi and S. Doi : Trans. of AFS **101** (1993) 225
[4] K. Shimizu, T. Noguchi : Wear **176** (1994) 255
[5] K. Shimizu, T. Noguchi, T. Kamada, H. Takasaki : Wear **198** (1996) 150
[6] I. Finnie : Wear **3** (1960) 87
[7] J. G. A. Bitter : Wear **6** (1963) 5
[8] I. M. Hutchings and R. E. Winter : Wear **27** (1974) 121
[9] 吉野達治：エロージョンとコロージョン（東京裳華房株式会社）(1987) 146
[10] I. Finnie : Wear **19** (1972) 81
[11] R. E. Winter and I. M. Hutchings : Wear **29** (1974) 181
[12] I. M. Hutchings : Wear **70** (1981) 269
[13] G. Sundararajan and P. G. shewmon : Wear **84** (1983) 237
[14] Kazumichi Shimizu, Xinbayaer, Tadashi Momono, Hideto Matsumoto, Yoshiyuki Maeda, Kiyosuke Sugawara : Proceedings of the 2nd Japan-Korea workshop for young foundry engineers, Jeju island, Korea (The Korean Foundrymen's Society) (2005) 116
[15] D. G. Rickerby and N. H. Macmillan : Wear **79** (1982) 171
[16] A. K. Cousens and I. M. Hutchings : Wear **88** (1983) 335
[17] Ambrish Misra and Lain Finnie : Wear **65** (1981) 359

研究論文

硬質炭化物鋳ぐるみ鋳鉄のエロージョン摩耗特性

清水一道*　　　新巴雅尔*
松元秀人**　　　前田善之**

Research Article
J. JFS, Vol. 80 No, 8 (2008) pp. 457〜465

Erosive Wear Properties of Hard Carbides Cast-in Insertion Cast Irons

Kazumichi Shimizu*, Xinba Yaer*,
Hideto Matsumoto** and Yoshiyuki Maeda**

The erosion performance and micro-mechanisms of the erosion of WC enhanced cast irons are presented. Flake graphite cast iron (FC) and spheroidal carbide cast iron (SCI-VMn) were used as base metals to test the validity of this method. Samples were tested using an air shot blast machine. Their erosive wear properties were assessed and reaction layers also analyzed.

For WC enhanced FC, it was confirmed that inserting WC near the test surface increases the erosive wear resistance of the base metal. However, for WC enhanced SCI-VMn, the erosion rate did not seem to improve by inserting WC.

Accordingly, the hardness of the inserted WC is thought to have a direct effect on wear amounts. It was also confirmed that if the formed reaction layer grows excessively, hardness of WC decreases and becomes lower than that of impact particles, which may cause the erosion rate to increase.

Keywords : erosive wear, erosion rate, inserted WC, reaction layer

1. 緒　　言

粉粒体が材料表面に繰り返し衝突することで材料の表面が損傷，除去される現象をエロージョンと呼ぶ．特に固気二相流での粉粒体のエロージョン摩耗はサンドエロージョンと言われている．このような現象は，主に輸送用のパイプベント，ベルホッパー，バルブ，ヘリコプターのブレード，タービンブレード，ファンなどにおいて大きな問題となっている[1〜5]．また，粉鉱石，微粉炭などの粉粒体の輸送時にも多くの損傷が起こっている．このような配管系が摩耗損傷によって孔があくようなことがあれば，重大な事故を招く恐れがある．前述のような管路のエロージョンは配管の内部が摩耗するため，外観からは判断できず，定期的なメンテナンスでの部品交換や減肉部分の肉盛り溶接などの処置により未然に事故を回避しているが，安全性を高めるため，及びコスト低減のためにも耐摩耗材料の開発は，解決すべき重要な問題となっている[2]．

今日，最も多く用いられている構造材料は鉄鋼材料である．配管系の部品として，軟鋼，炭素鋼，ステンレス鋼や各種の鋳鉄といった様々な鉄鋼材料が使用されている．以前の研究により，延性材料は衝突角20〜30°付近で摩耗量が多く，脆性材料は衝突角80〜90°付近で最大摩耗を取るという衝突角度依存性があり[6]，軟鋼などの延性材料を配管材料として使用するとき，どうしても30°付近で摩耗してしまうことが示されている．そこで実際に粉粒体の輸送配管などの重要部分には耐摩耗性の良好な材料を用い，摩耗による部品破壊を防がなければならない．

これまでの著者らの研究において，鋳鉄にバナジウムを約10 mass％添加し，基地中にバナジウムの球状炭化物を晶出した球状炭化物鋳鉄（SCI:Spheroidal Carbide cast Iron）が高い耐エロージョン摩耗特性を得た．炭化物が球状に晶出すると，応力分散が生じ，耐エロージョン摩耗材料には良好である．また，Mnを13 mass％添加させた高マンガン球状炭化物鋳鉄（SCI-VMn）はマトリックスがオーステナイトであるため，エロージョン摩耗進行時の衝突粒子により，マトリックスがマルテンサイトに加工誘起変態したことで硬さが著しく上昇し，耐エロージョン摩耗特性が向上した[1, 4]．ところが，実際の現場で製作するとき，温度管理によるVCの均一分散が非常に困難であり，または，VCが微細化し，粒径が一定しないのが現状である[6]．

平成19年11月12日　原稿受理
 *　室蘭工業大学材料物性工学科　　Department of material science and engineering, Muroran institute of technology
**　三共合金鋳造所　　　Sankyo. Co. ltd

その微細化部分がエロージョン摩耗に弱くなり，材料の耐エロージョン摩耗特性を低下させる．

そこで本研究では優れた耐エロージョン摩耗特性を示したSCIを基地組織として，WC粒子を鋳ぐるむことにより，更なる耐エロージョン摩耗特性を示す材料の開発を目的とした．

鋳ぐるみ鋳造法[7-10]は，鋳型にほかの機能を持つ材料をあらかじめ固定し，溶融させた金属（湯）を鋳型に流し込み作製する複合材料加工方法の一つである．鋳ぐるみ鋳造法は，大きさ・形状が比較的自由であり，鋳込み材質に制限がなく，設計変更に容易に対応でき，母材の特徴を生かしたままでの強化が可能である．

2. 実験方法及び硬質炭化物粒子

供試材は，鋳ぐるみ鋳造法で加工した片状黒鉛鋳鉄のFC鋳ぐるみ材及び高マンガン球状炭化物鋳鉄のSCI-VMn鋳ぐるみ材である．SCI-VMnは，過去の研究[11, 12]において，基地中の残留オーステナイトがマルテンサイトに加工誘起変態することで良好な耐エロージョン摩耗特性が得られたことより選んだ．FC鋳ぐるみ材は比較材料である．そして，これらの耐エロージョン摩耗特性を評価するために，同じ湯を用いてそれぞれの母材を作製し，鋳ぐるみ鋳造法を施した試験片と比較し，耐エロージョン摩耗特性が改善されるかを検討した．**Table 1**に供試材の化学成分を，**Fig. 1**に母材の組織を示す．エロージョン摩耗試験には，市販の吸引式ブラストマシンを使用した．衝突粒子には，ケイ砂を用い，総循環質量を2 kgとした．一定の条件で行うため，ノズルからの固体粒子の衝突速度は100 m/sec. とし，実験は全て室温・大気中で行った．質量計測は600 sec.，1800 sec. 及び3600 sec．ごとに行い，実験時間は3600 secとした．固体粒子は実験経過に伴い損耗あるいは粉砕するため，試験の都度交換した．

摩耗質量は感度1 mgの電子天秤により，試験前後の重量差から求めた．しかしながら，本研究では密度の異なる材料を使用したため，その耐エロージョン摩耗特性を比較する場合，材料の摩耗損傷は重量減少量で比較するのではなく，体積減少量で比較する方が適当である．そこで，供試材の平均密度を用い，損傷速度[13]（Erosion rate, cm³/kg）で摩耗量を算出した．鋳ぐるみ材料は硬質炭化物粒子が流動せず鋳ぐみ鋳造法が施されたと想定し，平均密度を求め，損傷速度を算出した．**Fig. 2**に試験片の形状を示す．試験片の寸法は粉粒体が全て当たり，また鋳ぐるみの効果を見るために，適切な厚さが必要とされることから，50×50×20（mm）の平板状のものを用いた．

Fig. 3（a）（b）に硬質炭化物粒子及びその配置を示す．硬質炭化物粒子には平均粒径6 mm，ビッカース硬度1350 HV，形状は粒状のタングステンカーバイド（WC）を使用した．WCをこのまま鋳ぐるみを施すと，溶湯によって流動してしまうので，Fig. 3（b）に示すように硬質炭化物粒子を網の上に均一に配置した上で，注湯し作製した．

WCは非常に高硬度であり，引っ掻き摩耗やすべり摩耗の耐摩耗材として広く利用されていることから，WCを鋳ぐるむことで耐エロージョン摩耗特性の向上が期待できる．粒径については，過去の研究より，平均粒径2 mm，不定形のWCの鋳ぐるみを施し実験を行った[14]結果，硬質炭化物粒子が小さいため，母材が摩耗すると同時に剥離してしまい，摩耗量が増加した．そのことを踏まえ，粒子の形状を大きくし，平均粒径6 mmを使用した．形状を球形にし

Table 1　Chemical composition of specimens. (mass%)

specimen	C	Si	Mn	V	Ni
FC	3.24	1.79	0.26	-	0.26
SCI-VMn	2.7	0.87	13.0	13.0	0.91

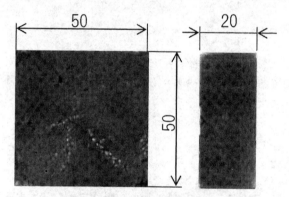

Fig. 2　Specimen dimensions. (mm)

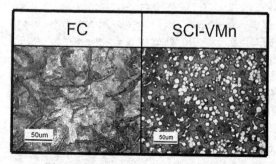

Fig. 1　Microstructure of base metals.

Fig. 3　WC for inserting and their distributions on test surface of specimens.

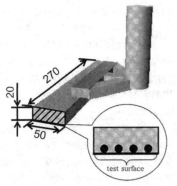

Fig. 4 Casting design for FC+WC. (mm)

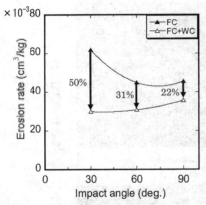

Fig. 5 Erosion rate vs. Impact angle in FC and FC+WC.

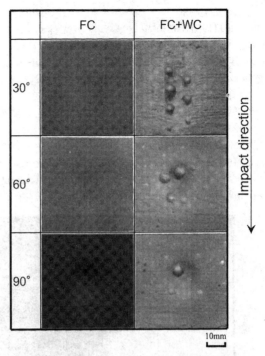

Fig. 6 Eroded surfaces of FC and FC+WC.

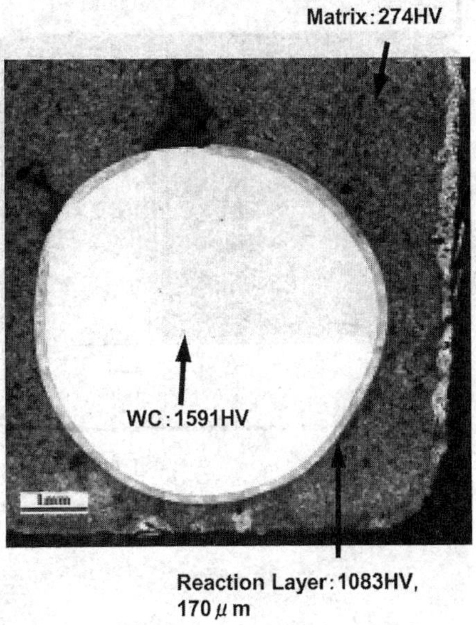

た理由については，基地組織内に球状炭化物が晶出していると応力分散により良好な結果が得られた[1]ことより選定した．衝突粒子には，不定形の宇部6号ケイ砂を使用した．

3. 実験結果及び考察

3.1 FC及びFC鋳ぐるみ材

3.1.1 鋳造方案

鋳造方案について，平板鋳ぐるみ鋳造方案を用いた．Fig. 4にFC及びFCの鋳ぐるみ材の鋳造方案を示す．

3.1.2 エロージョン摩耗試験結果と摩耗面マクロ観察

Fig. 5にFC及びFC鋳ぐるみ材（FC+WC）のエロージョン摩耗試験結果を示す．FC+WC材は母材FCと比較した場合，全ての角度においてエロージョン摩耗が抑えられており，耐エロージョン摩耗特性が向上したことがわかる．具体的に損傷速度が母材と比べ，30°では50%，60°では31%，90°では22%程度抑えられた．この結果から，WCを鋳ぐるむことで，耐エロージョン摩耗特性が顕著に向上することが確認された．また，FCでは低角度の切削摩耗で摩耗のピークを示したが，WCを鋳ぐるむことにより，角度依存性も改善された．

FC鋳ぐるみ材において，試験後のエロージョン摩耗面をデジタルカメラにより接写し，また摩耗面マクロ観察を行い，摩耗面でのWCの母材との接合性を調査した．その結果をFig. 6に示す．FC鋳ぐるみ材の摩耗面には，硬質

Matrix：274HV

WC：1591HV

Reaction Layer：1083HV, 170μm

Fig. 7 Thickness and hardness of bonded section in FC+WC.

炭化物粒子が粒状で現れた．90°では垂直方向に粉粒体が衝突し摩耗するため，硬質炭化物粒子の上部のみが摩耗面に現れたが，30°では硬質炭化物粒子の側面も現れている．60°の場合では30°及び90°の両方の特徴が現れている．摩耗面に現れる硬質炭化物粒子の表面積が大きいほど，耐エロージョン摩耗特性の向上率が高くなっていることがわかる．

171

3.1.3 反応層

WCと母材の間はどうなっているのかを調査するため，WC粒子付近の組織観察をした．その結果を **Fig. 7** に示す．WCの外周囲に均一な反応層が形成されたことがわかる．反応層の厚さは0.17 mm，また母材の硬さ274 HVに対して，反応層の硬さは1083 HVとなり非常に硬くできていることが観察された．硬くて均一な反応層ができることよりエロージョン摩耗が抑えられたと考察できる．

3.1.4 EPMA面分析

反応層成分を同定するために，電子プローブマイクロアナライザー（EPMA）分析を行った．**Fig. 8** にFC+WC材のEPMA面分析結果を示す．反応層にはWを主成分に少量のFe，Cが含まれていることをわかる．したがって，反応層はW，Fe及びCの化合物であることが確認された．

以上のことより，母材にFCより良好な耐エロージョン摩耗材料を用い，鋳ぐるみを施すことより更なる耐エロージョン摩耗特性を示す材料の開発を試みる．

3.2 SCI-VMn及びSCI-VMn鋳ぐるみ材

3.2.1 鋳造方案

SCI-VMn鋳ぐるみ材の鋳造方案を**Fig. 9**（(a)，(b)）に示す．反応層が確実にでき，50×50×20 mmの試験片が作製できるように，50×270×20 mm平板にて作製した試料を平板鋳ぐるみ材とし，50×270×250 mmのYブロック型にて作製した試料をYブロック鋳ぐるみ材とした．この図

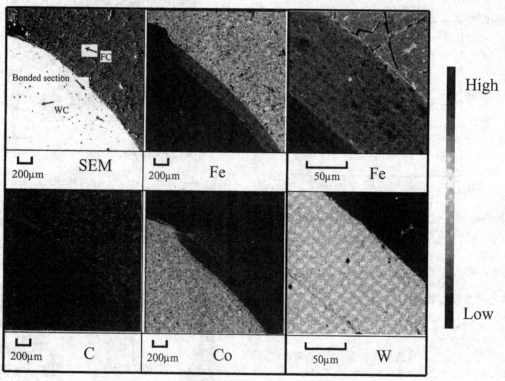

Fig. 8 EPMA surface analysis of FC+WC.

(a)   (b)

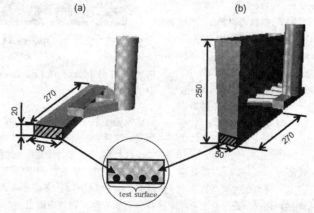

Fig. 9 Two kinds of casting design for cast-in insertion. ((a): plate type, (b): Y-block type) (mm)

に示すように，この鋳型の底部に硬質炭化物粒子を針金の上に均一に配置し，溶湯を鋳型に注湯し作製した．特にYブロック鋳ぐるみ材は，上からの熱供給が長いため，反応層が厚くできると予測できる．

3.2.2　エロージョン摩耗試験結果と摩耗面マクロ観察

Fig. 10にSCI-VMnの母材，平板鋳ぐるみ材（SCI-VMn+WC(P)）及びYブロック鋳ぐるみ材（SCI-VMn+WC(Y)）の実験結果を示す．SCI-VMnの平板鋳ぐるみ材では，全ての角度において摩耗をやや抑えることができたが，SCI-VMn+WC(Y)材は，まったく効果なく，むしろ母材

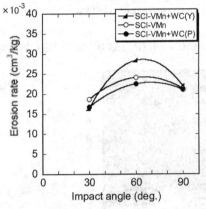

Fig. 10　Erosion rate vs. Impact angle in SCI-VMn, SCI-VMn+WC (P) and SCI-VMn+WC (Y).

よりも摩耗が増加した．そこで，WCの母材との鋳ぐるみ具合を明らかにするため，摩耗表面のマクロ写真を観察した結果を**Fig. 11**に示す．前述したようにSCI-VMnは1750℃の高温で注湯するため，WCを固定する網がすぐに解けることで，WCが元の位置から移動したか，または凝縮したため，WCの効果が出なかったことがわかった．

3.2.3　反応層

Fig. 12(a)にSCI-VMn+WC(P)材，(b)にSCI-VMn+WC(Y)材のそれぞれのエロージョン試験前の硬質炭化物粒子を示す．両案での硬質炭化物粒子には反応層の形成が確認された．反応層の厚さは鋳造方案により異なっている．SCI-VMn+WC(P)材で0.84 mm，SCI-VMn+WC(Y)材で1.56 mmとなった．Yブロック鋳ぐるみ材の反応層厚さは，平板鋳ぐるみ材の反応層より約2倍厚くなっていることにおいて，Yブロック鋳ぐるみ材の予測通り凝固時間が長く，より長い時間，溶湯から熱を供給されていることにより，反応層の形成が進行したと推察する．反応層が厚く形成されているにもかかわらず，摩耗量を抑えることができなかったのである．これについて，EPMA面分析及び硬さ分布にて考察する．

3.2.4　EPMA面分析

SCI-VMn+WC(P)材，SCI-VMn+WC(Y)では，EPMA分析による反応層成分は，**Fig. 13**及び**Fig. 14**に示すように，SCI-VMn+WCでは，反応層の外周部にはVが高濃度で検

173

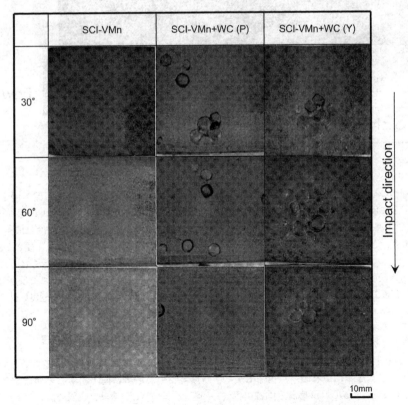

Fig. 11　Eroded surface of SCI-VMn, SCI-VMn+WC (P) and SCI-VMn+WC (Y).

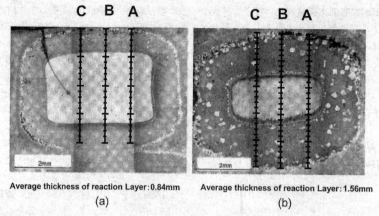

Fig. 12　Thickness and Hardness of bonded section in (a) SCI-VMn+WC (P) and (b) SCI-VMn+WC (Y).

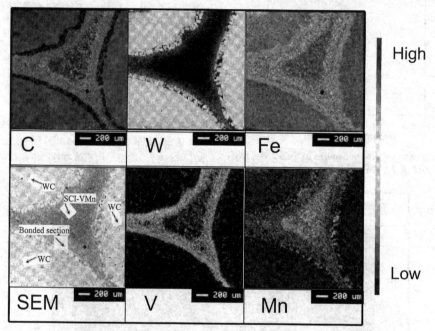

Fig. 13　EPMA surface analysis of SCI-VMn+WC (P).

出されるが，反応層への拡散は確認できない．Wは母材側に拡散しているが，主に反応層に含まれていることが明瞭である．したがって，反応層はW，Fe及びCの化合物であることが確認された．

3.2.5　硬さ測定

マイクロビッカース硬度計（VHN300）にて試験前後の硬度測定を行った．Fig. 15 (a)にSCI-VMn+WC(P)材，(b)にSCI-VMn+WC(Y) 材の母材－反応層－WCによる硬度分布を示す．鋳ぐるみを施す前のWC中心部は1350HVに対し，平板鋳ぐるみ材でのWCはFig. 12 (a)に示したようにA, B, C三箇所による平均硬さが1100HVとなり，反応層はWCの中心部より遠くなるにつれ硬度が低下している．Yブロック鋳ぐるみ材でのWCは反応層と同様に平均で800HVになり，平板鋳ぐるみ材と比べ低い値となって

いる．両案を比較すると，反応層の硬さがほぼ同等であるが，WCの硬さは元々の硬さより低くなり，特にYブロック鋳ぐるみ材では反応層が過度に成長したことで，WC内のWが反応層へ過度に拡散し，硬さが低下することで，粉粒体の硬さ（1030HV）より低くなったため摩耗が増加したことが明らかとなった．そこで反応層が過度に成長すると，逆にWCが軟らかくなり，良い効果が得られないと考察できる．

4.　硬質炭化物粒子の改良

以上のことより，WCを鋳ぐるむことで耐エロージョン摩耗特性が顕著に向上することが確認できたが，注湯の際，球状のWCが溶けた金属の流れにより移動してしまい，鋳ぐるみの効果に影響される．そこでWCの形状を改良

174

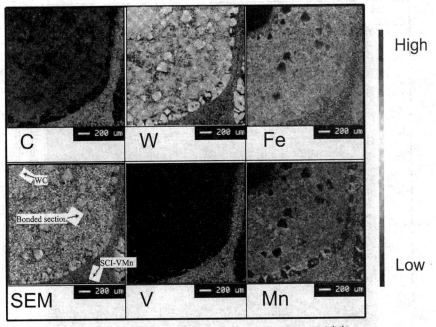

Fig. 14   EPMA surface analysis of SCI-VMn+WC(Y).

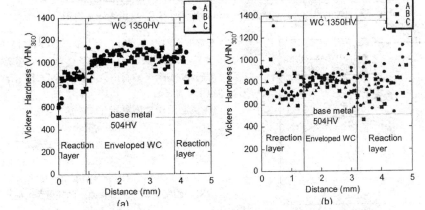

Fig. 15   Vickers hardness (VHN$_{300}$) distribution of reaction layer and enveloped WC for (a) SCI-VMn+WC (P) and (b) SCI-VMn+WC (Y).

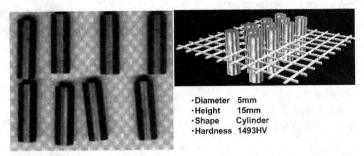

Fig. 16   Cylindrical WC for inserting and their distribution on test surface of specimens.

175

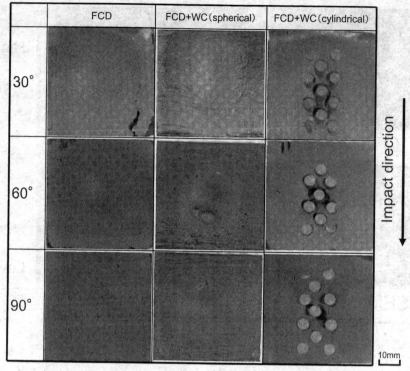

Fig. 17 Eroded surface of FCD, and cylindrical WC enhanced FCD.

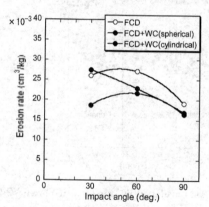

Fig. 18 Erosion rate vs. Impact angle in FCD, FCD+WC (spherical) and FCD+WC (cylindrical).

し，円柱状のWCを使用し，FCD母材に鋳ぐるみを施した．
Fig. 16 に円柱状のWC及びその配置を示す．

Fig. 17 に示すように，エロージョン試験後の表面マクロ写真から，円柱状のWCを使うことでWCが移動せず，良好に鋳ぐるまれていることがわかる．

Fig. 18 にエロージョン摩耗試験結果を示す．円柱状のWC鋳ぐるみ材のエロージョン特性が良くなっていることがわかる．

5. 結 言

本研究ではSCI系鋳鉄の耐エロージョン摩耗特性の向上を目的として，WCを鋳ぐるみ粒子として鋳ぐるみ鋳造法を施した材料を作製した．それらの材料に対し，エロージョン摩耗試験を行い，耐エロージョン摩耗特性について評価及び鋳ぐるみ材の反応層の解析を行った．その結言を以下にまとめる．

1) 表面近傍に硬質炭化物を鋳ぐるむことにより，耐エロージョン摩耗特性が向上する．

2) 硬質炭化物粒子と母材の間に反応層が形成することで，強固に鋳ぐるむことができ，摩耗による剥離が生じない．しかしながら，反応層が過度に成長すると，反応層及びWCの硬度が低下するため，摩耗量が増加する．

謝辞

最後に，本研究の試験片作製に（株）三共合金鋳造所のご協力を得た，ここに感謝の意を表する．また，本研究に協力して頂いた室蘭工業大学大学院修士坂本卓君（現；㈱前川製作所に所属）に御礼申し上げます．

参考文献

[1] Xinbayaer, Kazumichi Shimizu, et al: J. JFS **78** (2006) 510

[2] 清水一道，鉄鋼材料のエロージョン摩耗特性評価，学位論文（2001）79

[3] 吉野達治，エロージョンとコロージョン，（東京裳華房株式会社（1987）19

[4] 球相材料研究会，球状炭化物鋳鉄，（日刊工業新聞社）（2006）86

[5] Yoshinori Isomoto, Miyuki Nishimura, et al: Zairyo-to-Kankyo **48** (1999) 355

[6] Shigenori Nishiuchi, Tadashi Kitsudo and Hideto Matsumoto: J. JFS **79** (2007) 133

[7] Setsuo Aso, Hiroyuki Ike, Nobuo Konishi: J. JFS **76** (2004) 511

[8] Noritaka Horikawa, et al: J. JFS **75** (2003) 95

[9] Toru Noguchi, Noritaka Horikawa, Kazunori Asano: J. JFS **79** (2007) 206

[10] Noritaka Horikawa, et al: J.JFS **73** (2001) 668

[11] Kazumichi Shimizu, Xinbayaer, et al: Proceeding of the 2nd JSME/ASME international conference on materials and processing and the 13th JSME materials and processing conference (2005) (p3) 1

[12] Xinbayaer, Kazumichi Shimizu, et al: Proceeding of the 3rd Asia International Conference on Tribology (2006) 23

[13] I. Finnie: Wear **3** (1960) 87

[14] Setsuo Aso, Masashi Nakanishi, et al: J. JFS **73** (2001) 155

Erosion characteristic of white cast iron with spheroidal carbides

Kazumichi SHIMIZU (Muroran Institute Technology)
Xinbayaer (Muroran Institute Technology)
Hideto MATSUMOTO (sankyo.Co.Ltd)
Yoshiyuki MAEDA (sankyo.Co.Ltd)
Tadashi MOMONO (Muroran Institute Technology)

Abstract

Erosive wear tests were performed on white cast iron(WCI), ferritic ductile iron (FDI), white cast iron with spheroidal carbides (KSH-W) and high manganese cast iron with spheroidal carbides (KSH-M) using a shot blast machine. Erosion damage was measured by the removed material volume at impact angle between 10 deg. and 90 deg. The surface metal flow in vertical sections was also observed. The mechanism of erosive wear, the effect of impact angle, and differences in wear features of specimens were discussed.

Experiment showed that, after initial stage, the erode volume increases almost linearly with blasting time in WCI, FDI and KSH-W. The erosion rate for KSH-W is about 1/2-1/3 of that for WCI and FDI. The surface hardness of eroded KSH-W specimens increased from the initial HV600 to HV 750 after 3600 sec. of blasting. It showed that, the surface structure transformed retained austenite to martensite, hardening the surface and lowering the erosion rate.

It was shown that KSH-W has excellent erosion resistance and it is expected to find wide applications as a wear resistant material.

Key words: erosive wear, spheroidal carbides, erode volume, wear resistant

178

1. INTRODUCTION

In recent years, there happened big accidents resulting from erosion wear phenomena according to the news relevant to it. Take the damage of pipe in Mihama power station for example, the thing that the wall of pipe was thinned down by erosion-corrosion wear result in the accident. Similarly, the erosion phenomena would be serious problems mainly at bend pipe, valve, turbine blade, and fan in pneumatic conveying system, and then blade of helicopter, etc.

The representative instances of sand erosion are, in secondary refining in iron and steel plant and smelting reduction equipment, when the dispersed particles such as dust coal, powdered mineral etc were blew into melted pig iron and steel from tuyeres, the erosion damage occurred at the bend part of pipe. If these kinds of pipe system were eroded to make holes, there may result in serious problems because of the gas and dispersed particles ejections.

So there have to change the pipe when performs regular maintenance, to do build up welding on the wall of pipe to be thinned down, to avoid the accidents happened. Therefore, to prevent the accident from occurring, estimation of life span during erosion and development of wear resistant material will be an important research work to do.

The dominant factors that influence erosion are mechanical properties of target materials, especially, hardness of them. In addition, the hardness, the shapes and sizes of impact particles, impact speed and impact angle of particles also affect the removed amounts of materials and erosion characteristics of materials [9-12].

In this research, cast irons with spheroidal carbides that are expected to wise wear resistant materials were employed and other common structure materials such as FC, FCD were used for comparison.

2. EXPERIMENTAL DETAILS

2.1 Testing Machine

A shot blast machine was used to test the erosion of specimens in this research. The sketch of the testing machine is showed in Fig.1.

2.2 Materials

Spheroidal carbides cast irons selected for the present investigation were white cast iron with spheroidal carbides (KSH-W) and high manganese cast iron with spheroidal carbides (KSH-M). And two kinds of tempered high chromium white cast iron (mass% of Cr is 26% and 17%; 26Cr (temper), 17Cr (temper)), flaky graphite cast iron (FC) and spheroidal graphite iron (FCD) were also used for comparison. The chemical compositions of the alloys are given in Table 1.

2.3 Metallography

Objective materials used in this research were characterized by optical microscope. The specimens measuring 10mm×15mm were used for metallographic processing. The grinding was finished with silicon carbide paper to 1000- -grit. The polishing was carried out on the diamond polishing machine to obtain a surface finish of 1μm. After that, the polished specimens of objective materials were etched by natal (the amount of HNO_3 was 5%). Finally, electronic microscope was used to evaluate the various phases in the microstructure of selected specimens. The metallographic structures of the alloys are shown in Fig.2.

2.4 Erosion Test

A shot blast machine was used to test the erosive wear of target materials in the present investigation. 2kg angular (irregularly shaped) silica sand with average diameter 410μm, Vickers hardness 1050Hv were used as impact particles. The impact particles were changed after each test because particles itself also were eroded and then its sizes will be changed. Specimens selected for the present investigation were cast iron with spheroidal carbides, high-chromium cast iron and some common structural materials. The specimens measuring 50mm×50mm×10mm were used for erosion test. The specimens were mounted into the test stage directly below the nozzle with a vertical distance of 50mm from the end of nozzle to the test surface into the erosion test machine by changing their impingement angles respectively 30, 60, and 90 degrees. The examined air speed was 145m/s, and the particle feed rate was measured with about 4g/s. All the erosion tests were conducted at room temperature in 3600sec. Before and after the test the amounts of specimens were weighed with electronic scale to prepare for measurements of erosion rate. It is more proper to compare by volume decreases than by mass decreases when comparing the removed materials from specimens that have different densities. So the erosion rate was calculated from the erode volume using average densities of target materials [6-8].

3. RESULTS

3.1 Observation of Experimental Result

Fig.3 (a) shows the weight loss of KSH-W as a function of erosion time for irregularly shaped silica sand particles at incidence angle of 30, 60, and 90 degree. The figure reveals that for three angles, after initial stage (from results of relevant works for erosion there was evidence of a short induction period before any weight loss was observed [1-3]), the weight loss is approximately linear with erosion time.

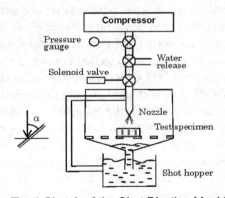

Fig. 1 Sketch of the Shot Blasting Machine

Table 1 The Chemical Compositions of the Alloys

	C	Si	M n	Cr	V	Ni	P	S	others
FC	3.24	1.79	0.26	-	-	0.06	0.01	0.01	Cu, Mg
FCD	3.74	2.16	0.3	-	-	0.02	0.01	0.01	Cu, Mg
KSH-M	2.92	0.57	12.9	-	11.9	0.95	-	-	-
KSH-W	2.79	0.96	0.54	-	12.7	3.06	-	-	-
26Cr	2.54	0.46	0.80	26.28	-	0.10	0.030	0.032	Cu, Mo
17Cr	2.97	0.50	0.72	17.28	-	-	0.030	0.024	Cu, Mo

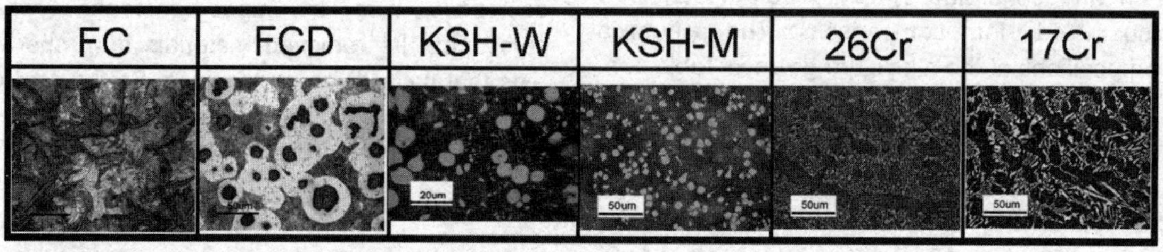

Fig. 2 Microstructure of the Alloys

180

Similar results were observed in WCI. Fig.3 (b) shows the results of 26Cr (temper).

As a preparation, the erosion test was perform--ed on KSH-M with steel grits. In the test with spherical steel grit, KSH-M has not been eroded and shows no clear peak. Furthermore, in the test with angular steel grit that have same average diameter and hardness as spherical one, the removed materials from surface is almost same as that in the test in spherical steel grit, From this point, angular silica sand that harder than steel grits was employed to be carried out erosion test on the materials. Fig.4 shows the result of erosion rate in KSH-M by both steel grit and silica sand. It can be clearly seen that the magnitude of erosion rate by silica sand is 2 orders of magnitude higher than that that by steel grit, so KSH-M almost were not eroded in erosion test with steel grit, whereas when the hard and angular silica sand was used as impact particles, there showed erosion in KSH-M. We draw conclusions that the removed

mass and angle independency also change dramatically if the impact particles were changed.

The erosion test was also performed on other materials with silica sand. The result is shown in Fig.5. Clearly seen from figure that flaky graphite cast iron (FC) is eroded mostly, KSH-W, as well as KSH-M, show best wear resistant properties. The expected wear resistant material of 26Cr (temper)and 17Cr (temper) that are very hard and have flake carbides in structure show peak at 60 deg., its wear resistant properties are superior to other common materials such as FC, FCD but are inferior to various spheriodal carbide cast irons.

3.2 Vickers Hardness of Specimens

In the erosion test, it is recognized that the work-hardening effect which resulted from impact of solid particles on material surface was happened. The harnesses of specimens before and after erosion test were measured through

Vickers hardness test.

Table 2 shows the results of them. From the differences of hardness before and after test in table, we can understand that there were happened the work-hardening effect on the material surface. In KSH-W and KSH-M which both have superior wear-resistant properties, the results of Vickers hardness were dramatically more increased than other materials after eroded. The surface hardness of eroded KSH-W specim--ens increased from the initial HV600 to HV 750 after 3600 sec. of blasting. Moreover that of eroded KSH-M reached to HV 804 from the initial hardness of HV 533.

While although the initial hardness of 26Cr (temper)and 17Cr (temper) were higher than FC and FCD, it almost not increased after erosion test and were more eroded than spheroidal carbides cast irons. From this fact, we can understand that it is true that erosion rate depend heavily on hardness of material, but compare with the initial hardness of material, the hardness after work-hardened is more important. It suggest that the residual austenite in matrix in KSH-W and KSH-M change to martensite by strain induced transformation, which results in increasing of hardness and sound erosive wear resistance.

181

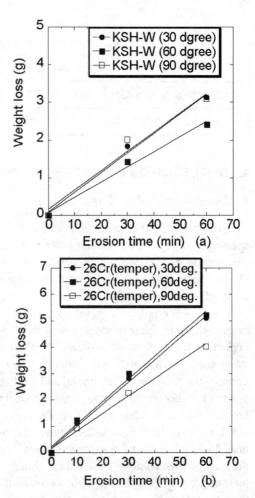

Fig.3 Weight loss vs. and Erosion rate in KSH-W (a), Hi-26Cr with Temper (b)

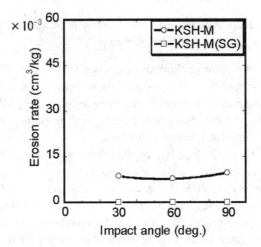

Fig.4 Erosion rate vs. Impact angle in KSH -M by Steel Grit and Silica sand particles

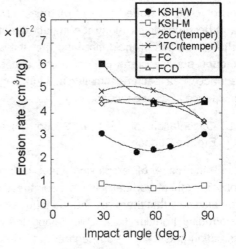

Fig. 5 Erosion rate vs. Impact angle in All specimens by Silica sand particles

Table 2   Vickers hardness of specimens both before and after erosion test

	KSH-W	KSH-M	FCD	FC	26Cr(temper)	17Cr(temper)
Hardness of before test	600	533	256	221	477	510
Hardness of after test	750	804	298	256	486	525

4. DISCUSSION ON ERODED MECHANISM

4.1 Observation of Wear Feature of Specimens

To make clear the erosion behaviors of selected specimens, the eroded surface were pictured by digital camera after finished the erosion test, Fig.6 shows the macrostructure of eroded surface by silica sand. In the previous works of our group, in the case of employing steel grit, it can be observed clear ripple pattern transverse to the impact direction of solid particles in 30 degree. Whereas in 60 degree, there have no clear ripple pattern, but indentation on upper part and ripple pattern just like in 30 degree from the center of eroded area to bottom. In 90 degree, there can not be observed ripple pattern but indentation only [4-5]; that is to say, there have clearly different erosive wear mechanisms in shallow and high degree that divided at about 60 degree. In the present research, however, in the circumsta--nce of silica sand, the macroscopic appearances of all specimens have not showed any ripples. It is observed that eroded area of materials are large and become longer along the impact direction at shallow angles of incidence and that the size of eroded surface areas is near round and eroded surface area became smaller at high angles of incidence in all specimens.

4.2 Observation of Vertical Section Near the Eroded Surface

In the case of employing steel grit, in the observation of vertical section near the surface, we also saw that there are formed tongue-shaped protruding portions that flowed along the impact direction plastically at 30 degree and 60 degree; whereas there are formed the compressed pilifo--rm protruding portions and indentations resulting from plastic deformation at 90 degree in the researches before [4-5]. On the other hand, in this work, employing silica sand, the vertical section near the surface is completely different from that using steel grit. Fig.7 shows the vertical section near surface of KSH-M employing silica sand. We observed from these photos that material was scraped away from surface by cutting and formed protrusion, at the meantime, the spheroidal carbide was preserved in just the shape and the matrix surrounding it was scraped at the shallow angle of 30 and 60 degree. So we can understand from this fact that the spheroidal carbide is very hard and its adhesion on matrix is also wise. In the case of 90 degree, because only compressive force stress on the material surface, the spheroidal carbides were pushed into matrix slightly; and also confirmed that the spheroidal carbides not only get off from material as it is but also break away from material but left residual into the eroded surface.

4.3 Correlations Between Rigidity of Material and Erosion rate for Specimens

We have confirmed erosive wear is relevant to the hardness of material but it is clear that hardly is hardness enough for preventing wear. We investigated the other mechanical properties of material in addition to hardness.

Fig.8 showed the relationship between rigidity and erosion rate in some specimens. It can be clearly seen that the erosion rate get to small when rigidity, i.e. internal frictional force become large. As for WCI although the rigidity of it could not have been measured, it still can guess that because of the lower rigidity of it, erosion wear of

WCI was more and less heavy in spite of higher hardness. So we consider that the erosive wear resistance of KSH-M was wise because of its higher rigidity although the initial hardness of it was lower (533HV).

In conclusion, it is to be desired that material which has high rigidity will be made when it comes to develop the erosive wear resistant material.

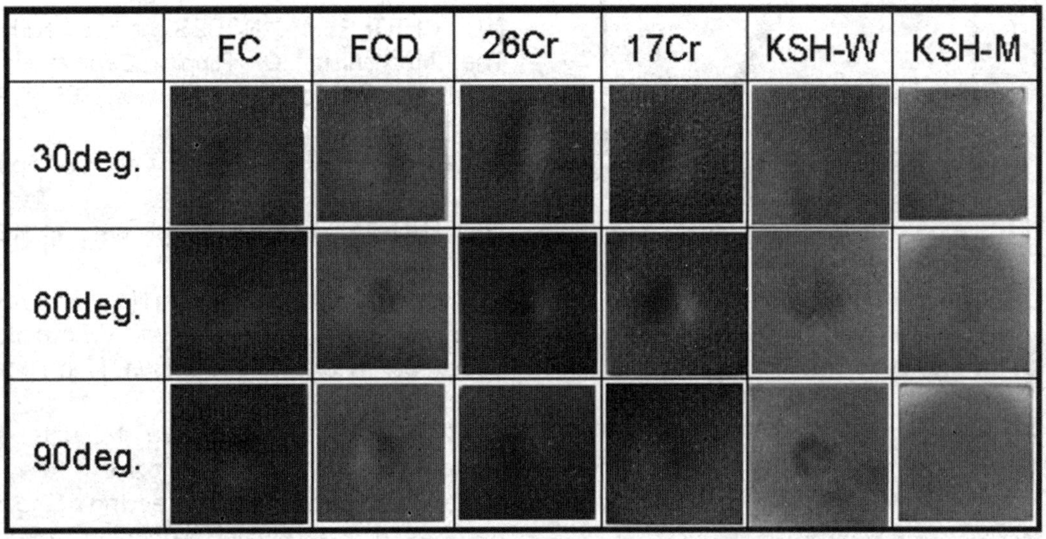

FC
FCD
26Cr
17Cr
KSH-W
KSH-M
30deg.
60deg.
90deg.

Fig. 6 The Macrostructure of Eroded surface in All selected specimens

30 deg.
60 deg.
90 deg.
50um
50um
50um

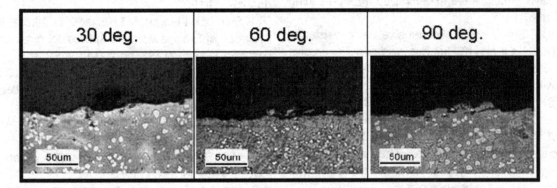

Fig. 7 The Vertical section near the Eroded surface in KSH-M at 30, 60 and 90 deg.

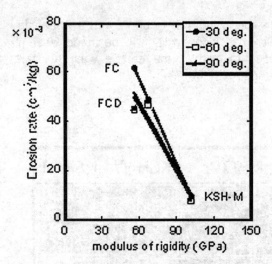

Fig. 8 Erosion rate vs. Modulus of rigidity in Specimens

5. CONCLUSIONS

Experiments were carried out to study the erosion behaviors and mechanisms of a series of iron and steel materials. The following conclusions were obtained

(1) Cast iron with spheroidal carbides has clearly superior erosive wear resistance compare with other materials. The reason is that stress concentration was dispersed because carbides in matrix structure are shperoidized through adding vanadium.

(2) KSH-W has higher erosive wear resistance than Hi-Cr, and KSH-M has most superior erosive wear resistance among the materials employed, owing to obvious work-hardening effect.

(3) Although it varies according to the shape and hardness of impact particles, the removed mass of materials is closely related to the rigidity of material and erosion rate. That is to say, the higher the rigidity of material is, the better the erosive wear resistance of it is.

(4) In erosion wear, the hardness after work-hardening is of more importance than the initial hardness. Therefore, the hardness of material increase because of strain induced transformation effect to show wise erosive wear resistance.

ACKNOWLEDGEMENTS

The authors are grateful to the whole staffs of material processing study room, muroran institute of technology, for giving assistances during all experimental works.

REFERENCE

[1] G. CHRTER, M. J. NOBES and K. I. ARSHAK, The Mechanism Of Ripple Generation on Sandblasted Ductile Solids, wear, 65 (1980) 151-174

[2] I. Finnie, Proc. 3rd U.S. Natl. Congr. on Applied Mechanics, 1958, p.527

[3] J. H. Neilson and A. Gilchrist, wear, 11 (1968) 111

[4] Kazumichi Shimizu and Toru Noguchi, Erosion Characteristics of Ductile Iron with Various Matrix Structures, Trans. of the Japan foundrymen's society, VOL. 13, NOV, 1994

[5] Kazumichi Shimizu, Toru Noguchi etc., fundamental study on erosive wear of austempered ductile iron, proceeding of the third east Asian international foundry symposium, pusan, Korea, 1-3 July 1992

[6] K.Shimizu, T.Noguchi, S.Doi, Basic Study on the Erosive Wear of Austempered Ductile Iron, Transactions of AFS, Vol.101, paper No.93-78, pp.225-229(1993)

[7] K.Shimizu, T.Noguchi, T.Kamada, H.Takasaki, Progress of Erosive Wear in Spheroidal graphite cast iron, Wear, Vol.198, pp.150-155(1996)

[8] K.Shimizu, T.Noguchi, T.Kamada, S.Doi, Basic Study of Erosion of Ductile Iron, Advanced Materials Research, Vol.4-5, pp.239-244 (1997)

[9] I. Finnie, Erosion of Surfaces by Solid Particles, Wear, 3, p87-103, 1960

[10] J.G.A.Bitter, A study of Erosion Phenomena PART1, Wear, 6, p5-21, 1963

[11] I.M.Hutchings and R.E.Winter, Particles Erosion of Ductile Metals: A Mechanism of Material Removal, Wear, 27, p121-128, 1974

[12] I.M.Hutchings, Mechanism of the Erosion of Metals by Solid Particles, ASTM STP664, p59-76, 1979

184